JOURNAL OF GLASS STUDIES

Scratch-engraved plate with two interlaced G's surmounted with a crown and four allegorical vignettes. Northern Europe, ca. 1640. Perhaps made for Gaston de France, the Duc d'Orleans, younger brother of Louis XIII. D. 48.8 cm. The Corning Museum of Glass *(No. 77.3.34).*

JOURNAL OF GLASS STUDIES

VOLUME 20 · 1978

THE CORNING MUSEUM OF GLASS
CORNING, NEW YORK
14830

The *Journal of Glass Studies* is published by The Corning Museum of Glass, Thomas S. Buechner, President; William C. Ughetta, Secretary; Van C. Campbell, Treasurer.

Subscriptions for the *Journal* should be sent to The Corning Museum of Glass, Corning, New York 14830. The price of the *Journal* is fifteen dollars a number plus postage and handling; one number is issued each year. The price of the *Index* to the *Journal of Glass Studies*, Vol. I–XV, 1959–1973, is eight dollars plus postage and handling.

Manuscripts, acquisition announcements, books and articles for the check list of recent publications should be addressed to the Editor, *Journal of Glass Studies*, The Corning Museum of Glass, Corning, New York 14830. All materials accepted for publication become the property of The Corning Museum of Glass.

PRINTED IN U.S.A.

Standard Book Number 87290–020–7
ISSN: 0022–4250
Library of Congress Catalog Card Number 59–12390

JOURNAL OF GLASS STUDIES

Contents

VOLUME 20 · 1978

I. GLASS FURNACES THROUGH THE AGES

ROBERT J. CHARLESTON

THE EARLIEST glassmaking complex known to us appears to be that at Tel-el-Amarna, the capital of Akhenaton, the heretic Egyptian king of the XVIIIth Dynasty (reigned 1375–1358 B.C.). Here an open-hearth system appears to have been employed, with shallow crucibles supported on refractory drums. No really high temperatures can have been attained, and the glass was probably worked in a pasty state, mainly by drawing out of threads and constant working of them by rolling on a flat slab, the forerunner of the modern "marver."[1] The glassy materials were probably subjected to a preliminary fritting process before founding. The first intimation of a more complex type of "furnace," probably reverberatory in character, comes to us from Assyria and is contained in a series of cuneiform tablets from Nineveh, dating from the seventh century B.C. Although the find of tablets was made in the royal library of Assurbanipal (668–637 B.C.), it seems likely that the texts derive from an older source or sources. Three types of furnace seem to be described, but unfortunately none of them in any detail. The first, "the *kuru*-kiln with four openings," appears to have been used for fritting; the word translated "openings" means literally "eyes," and it was prescribed that the material to be fritted should be placed between the eyes. The *kuru*-kiln had a "lower part" or "fireplace," so it may be supposed that the furnace was constructed, at least in its lower part, somewhat like a Roman pottery kiln, the main floor of which was pierced to allow the flames to ascend into the chamber above. That the ordinary "*kuru*-kiln" had an

---

This article was compiled a number of years ago for a publication which never materialized. It is now printed virtually as written then but has been brought up to date in some sections where new developments have rendered previous views obsolete and has been furnished with a critical apparatus of notes. It was written without benefit of A. Gasparetto's important article "A proposito dell' Officina vetraria Torcellana: Forni e sistemi di fusione antichi," in the *Journal of Glass Studies* (*JGS*), 9, 1967, pp. 50–75; and follows in its main lines the present writer's contribution on "Glass" to C. Singer, *et al.*, *A History of Technology*, 3, 1957, p. 206 ff.

1. First published by W. M. Flinders Petrie, *Tell-el-Amarna*, London, 1894, p. 25 ff. The "refractories" found on the Amarna site have been shown to have begun to vitrify at 1100°, and it seems unlikely that temperatures in excess of this could have been attained in this workshop. (W. E. S. Turner, "Studies in Ancient Glass and Glass-making Processes. Part I. Crucibles and Melting Temperatures Employed in Ancient Egypt at about 1370 B.C.," *Transactions of the Society of Glass Technology*, 38, 1954, pp. 436–444). See also Birgit Nolte, *Die Glasgefässe im alten Ägypten*, Berlin, 1968, pp. 22–33. Possible methods of working the glass made in these primitive furnaces have been suggested by Dominick Labino, "The Egyptian Sand-core Technique: a New Interpretation," *Journal of Glass Studies*, 8, 1966, p. 124 ff.

upper chamber, however, seems unlikely, since a second type of "*kuru*-kiln" is differentiated as "with a chamber." This type of furnace had a compartment ("chamber") which could be closed by means of a "door," and was evidently used for founding the glass. This was presumably a reverberatory furnace, but there is no indication of the path by which the heat rose within it. A third furnace, the "*atunu*-kiln," appears to have been used in the case of prolonged firings lasting up to a week.[2]

Whereas some glassmaking sites are known from Roman times, unfortunately none offers any structure which can be interpreted to give any idea of what the Roman furnace looked like. The only evidence on this point appears to be the representation, on a clay lamp attributed to the first century A.D., of what may reasonably be taken to be two glassworkers at a furnace. The details are far from clear, but it seems evident that the furnace was in at least two tiers, presumably a stoke-hole below (represented by the filling hole of the lamp itself) and a chamber above.[3]

Paradoxically enough, it is only with the Dark Ages that we begin to obtain evidence to indicate what early glassmaking furnaces were like. Under the unifying influence of the Roman Empire the practices of glassmaking in the different parts of the Empire appear to have been much the same. With the breakdown of the Empire, however, a cleavage seems to have developed between the practice of the North and that of the South, perhaps caused by the increasing difficulty experienced in the North of obtaining the soda ash which was a Mediterranean specialty, and a consequent reliance on the ashes of indigenous plants.[4] This cleavage appears to be reflected in the forms of the furnace.

I. The "Southern" Furnace

Our first reasonably full description of a "Southern" furnace appears in a Syrian manuscript in the British Museum which apparently cannot be dated earlier than the ninth century A.D. It runs as follows: "The furnace of the glassmakers should have six compartments, of

---

2. A. Leo Oppenheim, *et al.*, *Glass and Glassmaking in Ancient Mesopotamia*, Corning, 1970, pp. 32–33, 35, 40–41, 44–45, 48, 52, 58, 69 ff. An experiment to reconstruct a glass in the laboratory, by following the "recipes" of these texts, produced a satisfactory glass by means of fritting the ingredients at 920° and founding the frit at 1100° for sixteen hours (Robert H. Brill, *ibid.*, pp. 112–113). These temperatures may approximate those achieved in the Assyrian or Babylonian furnaces. The difficulties to be encountered in working with impure materials, however, are set out by W. E. S. Turner, "Studies in Ancient Glasses and Glass-making Processes, Part V. Raw Materials and Melting Processes," *Transactions of the Society of Glass Technology*, 40, 1956, p. 270 ff. For an earlier view of the whole problem of the Assyrian furnaces, see W. A. Thorpe, "Codes of Work in Glass History," *Journal of the Royal Society of Arts*, 1948, pp. 460–477; R. J. Forbes, *Studies in Ancient Technology*, 5, Leiden, 1966, pp. 141–143.

3. Michael Abramić, "Eine Römische Lampe mit Darstellung des Glasblasens," *Bonner Jahrbücher*, Hft. 159, 1959, p. 149 ff., pl. 27. The most revealing structural remains of identified Roman glass furnaces are those at Eigelstein, near Cologne, which represent the lower courses of both circular and rectangular structures, somewhat separated from each other, but without any indication of their relationship (O. Doppelfeld, "Die Kölner Glasöfen vom Eigelstein," Paper 236 read to the International Commission on Glass [ICG] conference, Brussels, 1965; *idem*, *Römisches und Fränkisches Glas in Köln*, Cologne, 1966, pp. 11–16). The putative glassmaking sites so far identified in England—Wilderspool (T. May, *Warrington's Roman Remains*, Warrington, 1904, p. 37 ff.), Caistor-by-Norwich (D. Atkinson, "Caistor Excavations," *Norfolk Archaeology*, 24, 1929, p. 108 ff.), Colchester (unpublished), and Mancetter (unpublished)—present no coherent picture of what form Roman glass furnaces in Britain assumed. The same is true of Gaul (see D. B. Harden, "Glass-making Centres and the Spread of Glass-making from the 1st to the 4th Century A.D.," *Annales du 1er Congrès des JIDV*, Liège, n.d., 1959, pp. 52–53, to which may be added H. Bievelet, "Fragment de sole d'un 'four à verrier' d'époque romaine découvert à Bavai," Paper 237 read to the ICG conference, Brussels, 1965). Excavations in Israel suggest that tank furnaces may have been used in the general "Syrian" area (Gladys D. Weinberg, "Roman Glass Factories in Galilee," Museum Haaretz, *Bulletin*, 10, 1968, pp. 49–50). Since this type of furnace never seems to have spread to Europe, which is the main focus of the present article, the details will not be discussed here.

4. See D. B. Harden, "Ancient Glass, III: Post Roman," *Archaeological Journal*, 128, 1972, p. 87.

which three are disposed in stories one above the other . . . the lower compartment should be deep, in it is the fire; that of the middle storey has an opening in front of the central chambers—these last should be equal, disposed on the sides and not in the centre [?], so that the fire from below may rise towards the central region where the glass is and heat and melt the materials. The upper compartment, which is vaulted, is arranged so as uniformly to roof over the middle storey; it is used to cool the vessels after their manufacture."[5] Not all the details of this account are clear, but the essentials are that the furnace is in three stories, with a fire chamber at the bottom, a central chamber into which the heat rises to melt the glass, and a vaulted upper compartment in which the glass may cool. This arrangement may be seen in the earliest representation of a glass furnace which has come down to us (Fig. 1). This is taken from a manuscript of 1023 in the library at Monte Cassino, a text of the work *De Universo* by Rabanus Maurus. Again, the details are not unequivocally clear, but the artist appears to have attempted to represent a cylindrical structure, although the roof is shown as a simple tent shape. The essential features, however, are these—a single stoke-hole at the bottom, a middle chamber with multiple "glory holes" giving access to the glass pots, and an upper compartment in which a glass may be seen annealing. It has been suggested that this illustration has been copied from one in a much earlier manuscript and somewhat misunderstood by the illuminator in the process.[6] It is nevertheless one of the most useful and eloquent pieces of evidence which we have. If indeed the furnace represented was round in section, the illumination provides a most useful link, for on the one hand the later furnaces of this type were almost always circular, whilst on the other the excavated remains of a seventh/eighth-century glassmaking complex on the Venetian island of Torcello include the foundations of a circular

FIG. 1. *Illuminated miniature from the manuscript of Hrabanus Maurus,* De Universo, *datable to 1023 A.D. Abbey of Monte Cassino (Codex 132).*

structure which was probably a glass furnace.[7] On a glasshouse site at Corinth (Agora, South Center site), however, dating from the eleventh/twelfth century, the foundations of a square

---

5. E. Dillon, *Glass*, London, 1907, pp. 122–124. See also Gasparetto, "A Proposito . . .," pp. 54–55.

6. E. Schenk zu Schweinsberg ("La fabrication du verre d'après deux manuscrits enluminés du 'De Universo' de Raban Maur," *Cahiers de la Céramique*, 32, 1963, pp. 268–273), who points out that the illustration in this manuscript is probably derived from an earlier original, perhaps in a manuscript of the *Etymologies* of Isidore of Seville (d. 600). Gasparetto, "A Proposito . . .," pp. 55–56, suggests that the originals may go back as early as the fourth-fifth century A.D. He sees the main body of the furnace as rectangular on plan, but this seems unlikely from the elliptical rendering of the "glory holes" (which would certainly not in any case be made at the corners of a rectangular structure); furthermore, the tent-like upper story is carefully depicted as being adapted to the curved structure below.

7. A. Gasparetto, "Les Fouilles de Torcello et leur apport à l'histoire de la verrerie de la Vénétie dans le Haut Moyen Age," Paper 239, read to the conference of the ICG, Brussels, 1965; A. Gasparetto, "A Proposito . . .," pp. 50, 68–75; E. Tabaczynska, "Remarks on the origin of the Venetian glassmaking Centre," in (eds.) R. J. Charleston, Wendy Evans and A. E. Werner, *Studies in Glass History and Design*, Old Woking (n.d., 1969), pp. 20–23. The structures in this glassmaking complex, apparently four in number, included the lowest courses of one circular building. Although interpreted by Signor Gasparetto as a fritting furnace, this seems far more likely to have been the

furnace were unearthed which, from the absence of any ancillary furnace structures, were reasonably taken by the excavators to have been the lowest story of a three-tier arrangement.[8]

As the finds at Torcello suggest, the circular furnace (perhaps in three stories) was probably from the first the type used by the Venetian glassmakers who, from at latest the mid-fifteenth century, began to dominate the world markets with their superior crystal glass.[9] When we come to the sixteenth century, we are fortunate enough to have an eyewitness account of an Italian furnace. In 1508 a Swedish priest named Peder Månsson traveled to Rome in his clerical capacity, staying there until 1524. Whilst there he interested himself closely in the glass industry, then unknown in Scandinavia and therefore possessing a special fascination for a traveler from the North, and compiled an account of it in his *Glaskonst* (or *Art of Glass*). He describes the furnace as follows:

> The second furnace, in which the glass is to be founded, is more difficult to build. It must be entirely built and walled up with damp clay capable of resisting the fiercest heat, and must be in the middle of a wide, roomy house. You lay the foundation wall round in a circle, with a diameter of 12½ ft. At the point where the furnace mouth is to be below in the ground, you lay no wall foundation. This furnace must have three arches, one above another. The first and lowest arch must occupy the whole interior right up to the walls, and be not higher than 2½ ft. from the floor and 1 ft. 8 in. in diameter.[10] On this arch must be set the pots in which the glass frit is put. Then you arrange the wall outside so that it has six thin ribs, and between each set of ribs you reserve an opening, through which the glass mass is drawn out, worked, inserted, and handled. Directly in front of each opening a pot must be set in the furnace with the batch in such a way that the height exactly suits the opening. The second arch is made 4 ft. 2 in. high above the first, extending over the whole

---

main (founding) furnace. Later evidence (Biringuccio and the French *Encyclopédie*) suggests that the fritting furnace was rectangular and the founding furnace circular (but Agricola's fritting furnace is circular). This perhaps confirms the picture provided by the Roman site at Eigelstein (Cologne), where in successive phases the circular structures and the rectangular structures stood clear of each other (see p. 10 above: also n. 28 below). The Torcello circular structure stands in the middle of the complex, as would be natural for the working furnace, to which all the surrounding structures are ancillary. The spread there of broken pots, glass waste, etc., is probably the result of the clearance and disturbance of the site at the time of demolition. At Monte Lecco too, the glass finds and scoria were mainly found outside the area occupied by the furnace and its surrounding working space (see n. 8).

8. Gladys R. Davidson, "A Medieval Glass-Factory at Corinth," *American Journal of Archaeology*, 44, 1940, no. 3, p. 297 ff.: *idem, The Minor Finds, Corinth 12*, Princeton, 1952, pp. 83–90, 107–122, 144–145, 264–265. The original conclusions have been modified (Gladys Davidson Weinberg, "A Medieval Mystery: Byzantine Glass Production," *Journal of Glass Studies*, 17, 1975, p. 127 ff.) by later finds, notably some limestone blocks covered with glass drippings which seem to suggest that either small tanks or square containers were used in the glassmaking process. Finds of refractory pots with glass adhering, however, found in association with the Agora Northeast glassmaking complex (if such it is) suggest the more normal method of founding the metal. Fragments of glass found in the Agora South Center site have been reinterpreted as pontil wads and suggest that (unless they were cullet, which seems somewhat unlikely) glass was actually worked in the vicinity of this furnace (i.e., that it was not merely a fritting or annealing furnace). The fact that there is so little space surrounding the furnace structure (barely half a meter on three sides) is the strongest counter indication.

A furnace excavated at Monte Lecco, in the Apennines about 30 km from Genoa, and probably dating from the late fourteenth or early fifteenth century, showed clearly a circular structure with a central fire trench running between two roughly segmental solid "sieges." The fire trench was dug down into an ash pit in the ground at the front, and had an outlet at the back, presumably for the clearance of ashes (S. Fossati and T. Mannoni, "Lo Scavo della Vetreria Medievale di Monte Lecco," *Archeologia Medievale*, 2, 1975 p. 31 ff., particularly Figs. 5–7; see also T. Mannoni, "A Medieval Glasshouse in the Genoese Apennines, Italy," *Medieval Archaeology*, 16, 1972, pp. 143–145 and Fig. 43). The furnace has been tentatively reconstructed as a three-tier structure on the basis of a picture, perhaps dating from about 1590, in the Oratorio of S. Rocco at Altare (Fossati and Mannoni, *op. cit.*, Fig. 14, and M. Caligari and D. Moreno, "Manufattura Vetraria in Liguria tra XIV e XVII Secolo," *ibid.*, pp. 26–28). The absence of any structures in the immediate vicinity of the furnace suggests that all the processes (fritting, founding, and annealing) were carried on in the one furnace, and that this was therefore likely to be a three-tiered structure.

FIG. 2. *Drawing from a late fifteenth-century manuscript. Vatican Library (Chigi F. VIII 158 f.93v.). See note 9.*

FIG. 3. *Drawing from a late fifteenth-century manuscript. Vatican Library (Chigi F. VIII 188 f.191). See note 9.*

furnace, except for a round opening 10–15 in. in diameter in the middle of the arch. Round the top of the opening there must be a ledge, so that the glasses, when put there to cool, may not fall down into the furnace. The third and top arch extends over the whole furnace, and there must be 3 ft. 4 in. between it and the second arch, and three openings, 10 in. broad, through which the smoke discharges and the glasses are put in to cool. The furnace mouth below in the earth should be 1 ft. 8 in. broad. You stoke it with dry wood, the length of which corresponds to the inner breadth of the furnace; and for this purpose one digs out the earth below in front of the furnace mouth. . . .[11]

---

9. Two illuminations in manuscripts in the Vatican Library portray glassmakers at work (Figs. 2 and 3 here). The cruder of the two (Fig. 2) shows two glassblowers sitting on three-legged stools, while the *tiseur* attends to his stoke-hole. On the furnace is written unequivocally "fornax vitr." It is represented clearly as in three stories, the upper two slightly set back, and this detail is repeated in the second miniature (Fig. 3). Here, however, four out of the six ribs are visible, rising from a broader ledge just below the level of the *bocca*. This ledge no doubt furnished space for the marver. The *bocca* is shown as a round-topped arched opening, and the glassblower to the right holds his iron with his right hand in the circular *boccarella* (or "little mouth"—see quote, p. 12). In front is a circular aperture through which perhaps the *tiseur* may check the condition of his fire: alternatively, it may indicate in false perspective the central opening of the furnace (cf. Fig. 4 here, and p. 14).

10. This must mean "in width." "Vault" would perhaps give a clearer idea of the construction than "arch." The present translation is taken from S. E. Winbolt, *Wealden Glass*, Hove, 1933, pp. 77–79. The Swedish text was originally published by R. Geete, "Peder Månssons Skrifter," *Samlingar utgifna af Svenska Fornskrift-Sällskapet*, Stockholm, 1913–1915, pp. 557–566. See also O. Johannsen, "Peder Månssons Glaskunst . . .," *Sprechsaal, Keramik, etc.*, 65, 1932, Nr. 21, pp. 387–388; H. Maurach, "Glasschmelzöfen in Alter Zeit," *Glastechnische Berichte*, 12. Jafrg., Heft 8, 1934, pp. 265–274. Both give reconstructions of the furnace described by Månsson (see also Mogens Schlüter in T. Jexlev *et al.*, *Dansk Glas i Renaessancetid, 1550–1650*, Copenhagen, 1970, pp. 127–129, with reconstructions of both Månsson's and Biringuccio's furnaces). Månsson's text has more recently been reprinted by J. E. Anderbjörk in "Den äldsta svenska skildringen av glastillverkning," *Glasteknisk Tidskrift*, 3, 1953, pp. 2–11.

11. This suggests that there may have been a fire trench in this furnace, and this detail, together with the mention of digging out the earth in front of the furnace, strongly recalls the furnace at Monte Lecco (see n. 8).

FIG. 4. *Woodcut illustrating Vannoccio Biringuccio,* De la Pirotechnia, *Venice, 1540.*

A woodcut illustrating a glass furnace of this type appears in the *De la Pirotechnia* of Vannoccio Biringuccio, published in Venice in 1540 (Fig. 4). This shows a hive-shaped furnace with six external ribs, the stoke-hole to the front, the "teasehole" man carrying an armful of faggots with which to replenish it. The glassblowers sit to either side on three-legged stools. In front of each (better followed on the right-hand side of the cut) is a screen to protect the blower from the glare of the "glory-hole" (*bocca* or "mouth") through which the glass pots are reached. Projecting from the side of the furnace below is "a marble shelf" supported on an arch—the rudimentary form of the modern "marver," on which the glass drawn from the pot is rolled to smooth it as a preliminary to blowing. Biringuccio describes his furnace as follows:

> Now in order to complete the purification, a round furnace is made built of rough bricks made from a clay that does not melt or calcine with the fire. Its vault has a diameter of about four *braccia*[12] and a height of six *braccia.* It is arranged in this way. First a passage for the fire is made which leads the flames into the middle of the furnace; around the circle at the bottom a shelf ¾ *braccio* wide is made on which are to be placed the pots that hold the glass, and this must be about one *braccio* above the ground. Around this five or six well-made little arches are built as supports for the vault, and under these are made the little openings which allow one to look inside and to take out the glass for working at will. Then the vault is continued to cover the glass, and only in the middle is a little opening of a *palmo* or less left. Above this vault another vault is made which seals up and covers the whole; this is two *braccia* high above the first so that it completes the reverberatory furnace. This is the cooling chamber for the works when they have been made, for if they did not receive a certain tempering of air in this, all the vessels would break as soon as they were finished when they felt the cold. . . .[13]

By a curious convention, the illustrator of Biringuccio's work, in representing the top story, or annealing chamber of the furnace, has shown the holes through which the glasses are inserted to anneal as seen from the outside, but has cut away the central panel of the furnace wall to reveal on the inside the hole through which the heat ascends from the founding chamber to the annealing chamber above. This cut is indeed defective in many respects. It does not, for example, illustrate the arches made above the "siege" to accommodate the working openings. These arches were structurally necessary to support the vault above them when the glass pots were changed, as was necessary where an old pot cracked or otherwise became defective, and a new, preheated, pot had quickly to be substituted with a minimum disturbance to the general working of the furnace—an excessively hot and fatiguing task which was regarded as the worst chore of the glassmaker's *métier.* Biringuccio gives some indication of how this was done in the section of his work dealing with glass pots: "After six or eight months from the time when they were made, when you wish to put them in the furnace in order to begin work, that place which you left open under the arches is a quarter closed with a wall and only enough space is left to allow one of the said vessels to enter. . . ."[14] The aperture was then closed with

---

12. A *Braccio* = 2 *palmi* = 55.36 cm (22.98 in.)—see *loc. cit.* in note following, p. 475.

13. Vannoccio Biringuccio, *Pirotechnia,* translated by C. S. Smith and M. T. Gnudi, New York, 1942, p. 128.

14. *Ibid.,* p. 129.

clay "making two small holes from one large one so that the worker can take out the glass with his tube from whichever vessel he wishes in order to work it. In the other opening he keeps another iron tube so that it will be hot.[15] Outside in front of these openings there is a support made of a marble shelf placed on an arch. Above this shelf and in front of the opening for the glass a screen is made to serve as a protection for the eyes of the workers and to carry an iron support which holds up the tube. . . ." These hooks (*halsinelle*) or supports for the blowing iron or pontil are not shown in Biringuccio's woodcut, but may be seen in Fig. 5. Probably at a quite early date it was found convenient to incorporate the working opening in a slab of fired clay which could be removed and replaced when the pots were changed. Fragments of such slabs were found on the early seventeenth-century glassmaking site at Jamestown in Virginia, U.S.A.[16] The structure of these arched openings, not shown on Biringuccio's cut, is fortunately illustrated in the woodcuts to what is perhaps the most systematic account of glassmaking surviving to us from the sixteenth century—the twelfth chapter of Georgius Agricola's *De Re Metallica*, first published in Basel in 1556. In many particulars Agricola shows himself as dependent on Biringuccio, but the situation in glassmaking which he describes is far more complex, probably because in his time, in Saxony and elsewhere in Germany, the indigenous "Northern" tradition was being penetrated by the Venetian tradition. This point will be dealt with later on (see pp. 20–29 below). In the meantime, it suffices to notice his far more detailed woodcut showing the three-tier furnace (Fig. 5). Here one may observe not only in the lower picture the arched openings to the "siege" story of the furnace (but with square *bocca* and no *boccarella*), but in the sectional rendering an indication of the vault supporting the "siege" itself, the central holes allowing the passage of the heat (the upper one square

FIG. 5. *Woodcut illustrating G. Agricola,* De Re Metallica, *Basel, 1556, showing the three-tier "Second glass-furnace."*

15. The *boccarella*. A roughly circular refractory slab with a small hole in the center, found at Torcello, may have served this purpose (Gasparetto, "A Proposito . . . ," p. 75, Fig. 9); a comparable slab was found at Monte Lecco (Fossati and Mannoni, *op. cit.*, pp. 49–50, no. 20). The former, however, in view of the small size of this hole, may have been a "port-plug" for closing the *bocca* during founding, cf. American examples *JGS*, 10, 1968, p. 195, Fig. 4.

16. J. C. Harrington, *Glassmaking at Jamestown*, Richmond, Virginia, 1952, pp. 23–25; J. Paul Hudson, "Glassmaking at Jamestown—1608–9 and 1621–4," *The Iron Worker*, n.d., p. 15, ill. Parts of refractory frames for glory holes were found at the Stenhule glasshouse in Denmark (T. Jexlev, *et al.*, *op. cit.* in n. 46, Fig. 11.3) and were reported from the Nejsum site further north (*ibid.*, pp. 90–91).

FIG. 6. *Painting by G.M. Butteri in the* Studiolo *of Francesco I de' Medici, showing the Grand Duke's glassworks in Florence, ca. 1570. Palazzo Vecchio, Florence.*

instead of circular), and the arched opening allowing access to the annealing chamber.[17]

These somewhat dry illustrations are happily infused with life for us by a single sixteenth-century painting showing Italian glassmaking. This is a wall painting by G.M. Butteri decorating the *studiolo* of Francesco I de' Medici (Grand Duke of Tuscany 1574–87) in the Palazzo Vecchio at Florence (Fig. 6). It shows the Grand Duke's glasshouse, which is known to have been of Venetian type.[18] Here in the background may be seen the great glowing furnace, the "gaffers" seated on their three-legged stools before the glory holes, from the glare of which they are protected by the fire-clay screens described by Biringuccio, whilst to one side and slightly lower down are the *boccarelle* accommodating three or more irons. The master to the left warms-in his glass at the glory hole, resting his iron on the lowest of the *halsinelle*. Above the masters' heads glow the apertures leading to the annealing chamber, through one of which a servitor (to the left of the picture) is placing a finished glass to cool. To the right of the seated gaffer with his back to the onlooker may be seen the stoke-hole, toward which the tease-hole man, apparently stripped to the waist, brings a fresh bundle of faggots. Above the furnace is a framework of beams on which the wood for fuel was set to dry, with a resultant very grave risk of fire. When the Crutched Friars glasshouse in London burned down in 1575 it had "within it neere fortie thousand billits of wood."[19]

The furnace of Butteri's picture appears to be a developed form of that shown in Biringuccio's or Agricola's woodcuts. The ribs of the furnace,

17. In his text Agricola refers to "six arched openings," but his illustration would suggest that there were at least eight. A similar discrepancy appears in his description of the two-story "second furnace" (see p. 15), which is said to be strengthened on the outside with five ribs whereas "in the wall of the upper chamber between the ribs there should be eight windows . . ."; the illustration suggests that there would have been at least eight ribs (Fig. 6 here). The translations used here are Winbolt's (n. 10); see also *Georgius Agricola: De Re Metallica*, trans. H.C. and L.H. Hoover, New York, 1950.

18. G. Taddei, *L'Arte del Vetro in Firenze e nel suo Dominio*, Florence (1954), pp. 48–49, 115–117, and pl. IV. This glasshouse seems to have begun operations about 1568/9. See also L. Zecchin, "La Toscana e il Vetro Muranese" (Camera di Commercio Industria e Agricoltura) *Giornale Economico*, Venice, 1954, p. 17; *idem*, "Bortolo d'Alvise ai Tre Mori," *Vetro e silicati*, Anno XIII, no. 76, 1969.

19. W.A. Thorpe, *English Glass*, 3rd ed., London, 1961, p. 99. The framework on which the wood was stacked was called in Tuscan "matèo," but in Venetian dialect "caramàl" (Zecchin, "La Toscana . . .," p. 20. It may be seen also in the late seventeenth-century German engraving illustrated in Fig. 9 in Part II.

delineated as rather thin in Biringuccio and Agricola, have here become of considerable depth, at least at the base, thus providing a much wider working surface for the gaffer. This detail is seen again in seventeenth-century representations of Italian-type furnaces (Fig. 7). One curious feature of the Florence furnace is its asymmetrical form at the right-hand side, where the vault appears to project in an overhang. It seems possible that this may be the beginning of what later became the tunnel "lehr."[20] We may perhaps look for the beginnings of this development as far back as Biringuccio. His own Italian text is not easy to follow, but the French translation of 1556 by "Maistre" Jaques Vincent offers a somewhat easier interpretation. He writes: "this cooling-off is effected by a certain opening made on the left-hand side (Biringuccio says "at the back"), and this channel is shaped like a trumpet; from it all the cooled vessels are skillfully drawn by means of a long iron, one after the other, in three or four goes, until they reach the mouth and are taken outside."[21] This "trumpet-shaped opening" is perhaps the beginning of the lehr, and the man standing at the right of Biringuccio's woodcut (Fig. 4) is no doubt performing the office of moving the glasses along from hotter to cooler positions. Christopher Merrett in his translation of A. Neri's *L'Arte Vetraria* (*The Art of Glass*, 1662) writes:

> The *Leer* (made by *Agricola*, the third *furnace*, to anneal and cool the vessels, made as the second was to melt the Metall, and to keep it in fusion) comprehends two parts; the *tower* and *leer*. The *tower* is that part which lies directly above the melting *furnace* with a partition betwixt them, a foot thick, in the midst whereof, and in the same perpendicular with that of the second furnace, there's a round hole [*Imperat.* and *Agric.* make it square and small] through which the flame and heat passeth into the *tower*; this hole is called *Occhio* or *Lumella*, having an Iron ring encircling it called the *Cavalet* or *Crown*; on the floor or bottom of this *tower* the vessels fashioned by the M[rs] are set to anneal; it hath 2. *Boccas* or mouths, one opposite to the other, to put the Glasses in as soon as made,

FIG. 7. *Frontispiece to the 1669 Latin translation of A. Neri's* L'Arte Vetraria, *published in Amsterdam.*

20. There has been considerable discussion of the origin of this word, spelled variously "lehr," "lear," "lier," and "leer"—see W. E. S. Turner, "That curious word 'Lehr'," *Transactions of the Society of Glass Technology*, 33, 1949, pp. 278–288. The riddle seems to have been solved by L. Zecchin ("À propos de quelques termes de verrier de Murano présents dans la littérature technique du XVIIe siècle," Paper No. 234 read to the ICG conference, Brussels, 1965; and *idem*, "Terminologia Muranese nel Libro di Neri," *Vetro e silicati*, Anno IX, n. 54, 1965, pp. 21–24), the word(s) "l'era," "era" being the Muranese term for the annealing furnace.

21. *La Pyrotechnie . . . traduite d'Italien en François par feu Maistre Jaques Vincent*, pub. Claude Fremy, Paris, 1556, p. 90.

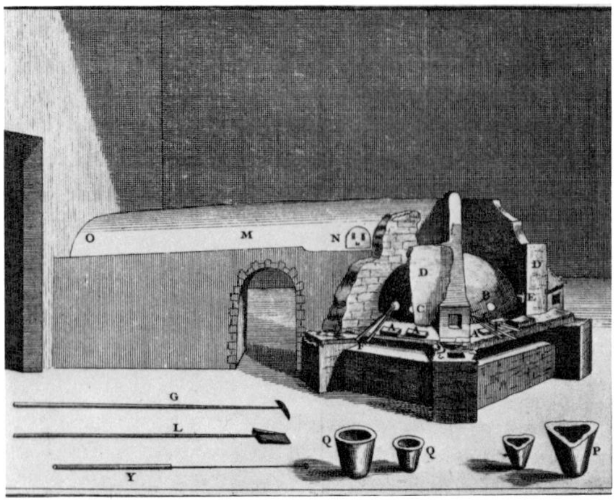

FIG. 8. *Plate IV from the French translation of A. Neri's* L'Arte Vetraria *(*Art de la Verrerie, *Paris, 1752), showing the "Amsterdam" furnace.*

FIG. 9. *Plate from Diderot's* Encyclopédie *showing "Verrerie en Bois, Coupe et Plans d'une petite Verrerie à pivette; et Coupe de la Cave à braise" (pl. vol. X, pl. 3).* oo *show the grills, composed of very short bars to span the firing channel:* ff *are the "fraches."*

FIG. 10. *Plate from Diderot's* Encyclopédie *showing "Verrerie en Bois, l'opération de retirer les Feraces et les transporter au Magasin" (pl. vol. X, pl. 22).*

FIG. 11. *Frontispiece to G.-L. Bianconi's* Due Lettere di Fisica, *Venice, 1746. Corning Museum of Glass. See note 24.*

taken with a Fork by the *Servitors*, and set on the floor of the *tower*, & after some time these Glasses are put into Iron pans [*Agric.* makes them of clay] call'd *Fraches*, which by degrees are drawn by the *Sarole man* all along the *Leer*, which is five or six yards long, that the Glasses may cool *Gradatim*, for when they are drawn to the end of the Leer they become cold. This *Leer* is continued to the *tower*, and arched all along about four foot wide and high within. The mouth thereof enters into a room, where the Glasses are taken out and set. This room they call the *Sarosel*. . . .[22]

This structure may be seen on the frontispiece of the 1669 Latin edition of Neri, published in Amsterdam (Fig. 7), and a similar engraving illustrates the 1752 French edition of the same work (Fig. 8).[23] In these furnaces, however, the glasses appear to have been put directly into

22. Christopher Merrett, *The Art of Glass*, London, 1662, pp. 243–244.

23. *Art de la Verrerie de Neri, Merrett et Kunckel . . . Traduits de l'Allemand par M.D. . . .*, Paris, 1752, pl. 4, entitled "Fourneau de Verrerie d'Amsterdam." The frontispiece, however, inscribed "ancien Fourneau de Verrerie à l'Italienne" is copied almost literally from the frontispiece of the Amsterdam editions of Neri. "M.D." was the Baron d'Holbach.

F I G. 12. *Model of a glass furnace, tin-glazed earthenware (*maiolica*). Italian; second half of eighteenth century. Crown Copyright. Science Museum, London (1934–188). See note 24.* L. 22 cm, W. 13.5 cm, H. (max) 17 cm.

F I G. 13. *Another view of the model shown in Fig. 12.*

the lehr through small doors (N of Fig. 8). The same feature appears in the great French *Encyclopédie* of Diderot and D'Alembert, in the volume of Plates dated 1772 (Fig. 9).[24] The same volume also gives a vivid view of the interior of the lehr seen from the Sarosel room, with the *fraches* moving in two lines down the tunnel (Fig. 10).[25] An improved form of tunnel lehr was invented by George Ensell at Coalbournbrook, in the Stourbridge glassmaking district, about 1780, but it is not certain what the improvement was.[26]

The Venetian-style glasshouse had one further subsidiary furnace. This was used for the preliminary roasting or "fritting" of the silica and ash. Biringuccio described this process as follows: "Then put all these things [that is, silica, soda, and manganese for decolorizing] mixed together into a reverberatory furnace made for this purpose, three *braccia* long, two wide, and one high, and apply enough of the strong flames of a wood fire by means of the reverberator, so that the composition is melted well and is converted all into one mass."[27] Peder Månsson confirms this general picture, adding the detail: "The mixture is often stirred

---

24. *Recueil de Planches sur les Sciences, les Arts Libéraux et les Arts Méchaniques . . . , Planches,* Tome X, Paris, 1772, pl. III. It may be opportune at this point to adduce two illustrations of Italian furnaces of the mid-eighteenth century. One of them (Fig. 11) is the frontispiece of *Due Lettere di Fisica al Signor Marchese Scipione Maffei . . .* by Gian-Lodovico Bianconi, published in Venice in 1746. Although it is difficult to envisage just how this furnace worked (it seems to have derived some of its features from Kunckel [see below, p. 26] although its basic principles are different, and it seems to have one story too many), the nature of its long lehr, with central archway, is unmistakable. The second piece of evidence is furnished by a tin-glazed pottery (*maiolica*) model of a furnace, in the Science Museum, London (Figs. 12 and 13). Here the long tunnel-lehr is clearly in evidence, with a glass visible in the "tower," apparently heated by the updraft from the founding furnace conveyed through a chimney which does not connect directly with the floor of the lehr. Figure 12 shows the stoke-hole of the main furnace, and to the left that (presumably) of the fritting furnace, with the end of the lehr above it. Figure 13 shows a second stoke-hole of the main furnace, and (presumably) the fritting platform incorporated into the same structure.

25. *Ibid.*, pl. XXII.

26. D. R. Guttery, *From Broad-Glass to Cut Crystal*, London, 1956, p. 97. The improvement may well have been the provision of a separately heated lehr, perhaps one with two tunnels, for large and small objects respectively, such as became standard in English glasshouses in the nineteenth century (see, e.g., D. Lardner, *The Cabinet Cyclopaedia*, London, 1832, pp. 158–159, 176–177). By the time Apsley Pellatt came to write *Curiosities of Glass Making* (London, 1849), the pans were moved mechanically, and the double lehr had become a quadruple one.

27. *Op. cit.* in note 13, pp. 127–128.

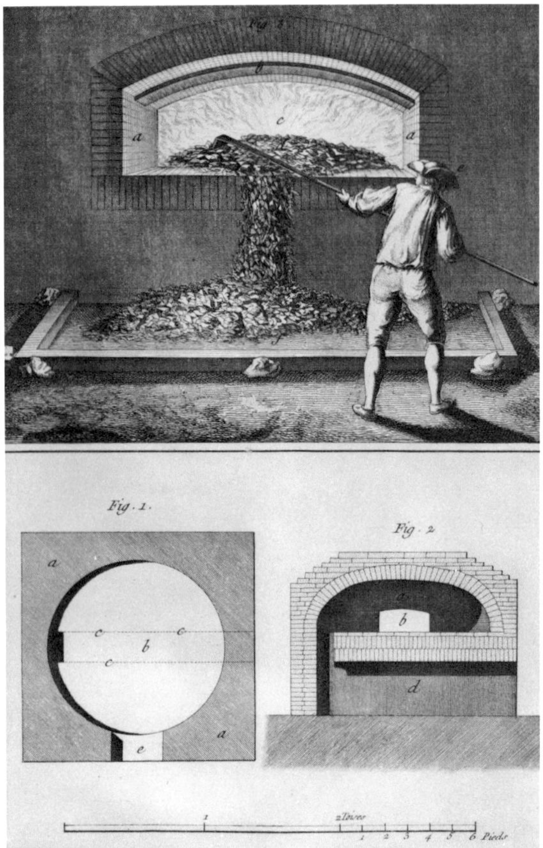

FIG. 14. *Plate from Diderot's* Encyclopédie *showing "Verrerie en bois, Plan et coupe de la Calcaisse, et l'Opération de retirer la fritte cuitte" (pl. vol. X, pl. 15).*

and turned round with an iron hook." It is probably a fritting furnace of this type that is to be seen to the right of Butteri's picture (Fig. 6).[28] No really explicit representation, however, is available to us until the time of the great *Encyclopédie*, which fortunately contains an engraving showing the furnace in plan and section and, for good measure, an illustration of the furnace-man at work raking the frit from the mouth of the furnace (Fig. 14; see note 28).

II. The "Northern" Furnace

There is no evidence to throw light on the "Northern" type of furnace nearly as early as that available to us for the "Southern" type. Our earliest source is Theophilus Presbyter, who in his *Schedula Diversarum Artium* (*Treatise on Divers Arts*) devoted a Book to glassmaking. The date of Theophilus has been much disputed, but the most recent view is that Theophilus was the Benedictine monk Roger of Helmarshausen and that his manuscript was compiled between 1110 and 1140 A.D.[29] If the identification is correct, the author was a practising metalworker who was personally able to carry out most of the techniques he described and who would certainly have described the glassmaking process with the insight of a craftsman. Theophilus describes the glass furnace as follows:

> If you have the intention of making glass, first cut many beechwood logs and dry them out. Then burn them all together in a clean place and carefully collect the ashes, taking care that you do not mix any earth or stones with them. After this build a furnace of stones and clay, fifteen feet long and ten feet wide, in this way.
>
> First, lay down foundations on each long side, one foot thick, and in between them make a firm, smooth, flat hearth with stones and clay. Mark off three equal parts and build a cross-wall separating one third from the other two. Then make a hole in each of the short sides through which fire and wood can be put in, and building the encircling wall up to a height of almost four feet, again make a firm, smooth, flat hearth over the whole area, and let the dividing wall rise a little above it. After this, in the larger section, make four holes through the hearth along one of the long sides, and four along the other. The work pots are to be placed in these. Then make two openings in the center through which the flame can rise. Now, as you build the encircling wall, make two square windows

28. Agricola's illustration of a fritting furnace also shows a round construction, although his verbal description seems to indicate an oblong structure ("Their *first furnace* should be arched over and resemble an oven. In its upper compartment, six feet long, four broad, and two high, the frit is cooked . . ."). The *Encyclopédie* version (*Planches*, pl. xv) is square on plan although the internal shape is circular. It has the advantage that it is fired at the side, for the greater convenience of the worker stirring the frit. Agricola's model would have burned his shins.

29. *Theophilus: the Various Arts*, translated by C. R. Dodwell, London, 1961, pp. xviii–xliv, and particularly p. xxxiii.

on each side, a span long and wide, one opposite [each] of the flame openings, through which the work pots and whatever is placed in them can be put in and taken out. In the smaller section also make an opening [for the flame] through the hearth close to the cross-wall, and a window, a span in size, near the short wall, through which whatever is necessary for the work can be put in and taken out.

When you have arranged everything like this, enclose the interior with an outer wall, so that the inside is in the shape of an arched vault, rising a little more than half a foot, and the top is made into a smooth, flat hearth, with a three-finger-high lip all around it, so that whatever work or implements are laid on top cannot fall off. This furnace is called the work furnace (Fig. 15).

(Chapter 2) Now build another furnace, ten feet long, eight feet wide, and four feet high. Then make a hole in one of the faces for putting in and taking out whatever is necessary. Inside, make a firm, smooth, flat hearth. This furnace is called the annealing furnace.

(Chapter 3) Now build a third furnace, six feet long, four feet wide, and three feet high, with a [fire] hole, a window and a hearth, as above. This furnace is called the furnace for spreading out and flattening.

The implements needed for this work are an iron (blow) pipe, two cubits long and as thick as your thumb, two pairs of tongs each hammered out of a single piece of iron, two long-handled iron ladles, and such other wooden and iron tools as you want.[30]

The treatise entitled *De Coloribus et Artibus Romanorum*, attributed to a certain Eraclius, contains chapters on glassmaking which have been added, probably in the twelfth or thirteenth century, to an existing manuscript.[31] These describe a tripartite furnace of which the largest section is in the center and is the founding and working furnace. This has one "glory hole" on either side, apparently with two pots to each, perhaps in the same manner as the Theophilus furnace. To the left of this should be a smaller furnace used for fritting and pot arching, and to the right a still smaller compartment presumably for annealing. The actual ground plan of these furnaces is not prescribed, and they

FIG. 15. *Model of Theophilus' furnace. Crown Copyright. Science Museum, London.*

may not necessarily have been rectangular.[32] That rectangularity was by no means the rule in practice seems to be shown by the evidence of a glass furnace excavated at Glastonbury in recent years. This appears to date from late Saxon times—say, between the eighth and the eleventh centuries A.D. Although not enough of the structure has survived to permit an exact

30. Translation quoted from *On Divers Arts: the Treatise of Theophilus*, translated by John G. Hawthorne and Cyril Stanley Smith, Chicago, 1963, pp. 49–52. The illustration here of Theophilus' furnace, made by the Science Museum, London, is probably wrong in showing four working-holes per side instead of two (see Hawthorne and Smith, p. 51, n. 2). The feature of the holes made in the siege to take pots is unique, and one is tempted to wonder whether Theophilus was not misled by seeing in a furnace the ring of glass left on the siege when a broken pot was removed. The reconstructions have always been made very trim and square; they were probably always somewhat more rough and ready in practice. It should be noted that here the small furnace is used for fritting, and a separate furnace for annealing. These procedures were often reversed, and sometimes the subsidiary furnace was used for both fritting and annealing. For the units of measurements used, see Hawthorne and Smith, pp. 49–51, n. 1.

31. For the dating of the Eraclius manuscript, see Dodwell, *op. cit.*, p. xiv.

32. For the text, see Mrs. Merrifield, *Original Treatises*, I, London, 1849, pp. 212–213. The fire trench down the middle of the whole furnace is clearly indicated, and it may be assumed that there were solid sieges to either side of it, rather than the improbably flimsy structure reproduced by H. Maurach, "Glasschmelzöfen in alter Zeit," *Glastechnische Berichte*, 12. Jahrg., Heft 8, 1934, Fig. 6 (after C. Friedrich).

FIG. 16. *Miniature from a manuscript of "Sir John Mandeville's Travels." Probably Bohemian; about 1420. British Library, London.*

reconstruction, the ground plan appears to have been oval.[33] Similarly, four furnaces attributed to the ninth century A.D. excavated at Nitra, in Slovakia, had oval ground plans. This has been interpreted to suggest that the Bohemian furnace was of a type "entirely different from the well-known description" in Theophilus.[34] It may well be, however, that the essential element in the Northern tradition was not so much the rectangularity of the ground plan as the fact that the main and subsidiary furnaces (or at least one of them) were on the same level and shared their heat either by having the same fire trench running the length of the composite furnace (as in Theophilus) or by transmitting the heat laterally from the main furnace to the subsidiary one, or by both. An arrangement of this sort may be seen in the famous illustration to Sir John Mandeville's *Tales*, in a manuscript in the British Museum (Fig. 16). This manuscript is considered to have been compiled in Bohemia about 1420. Here may be seen quite clearly a working furnace having two pots on either long side, a fire trench apparently running from end to end of the complex, and a subsidiary furnace being used in this case for annealing. The artist's rendering of the crown of the furnace would suggest an oval ground plan. An actual furnace of about this period excavated at Sklenarice (in the Semily district of the present Czechoslovakia) reveals an oval ground plan extended by two parallel lines of masonry, perhaps the original fire trench. Another structure hard by, this one apparently of rectangular form, has been interpreted as being the fritting furnace.[35] Yet another furnace of this general type has been excavated in Czechoslovakia at Ververi Bityska.[36] That the Bohemian furnace was not

33. Personal communication from Dr. D. B. Harden. See also D. B. Harden, "Ancient Glass, III: Post-Roman," p. 87, and sources quoted there.

34. D. Hejdová, "Glassmaking in the Archaeological Finds on the Territory of Czechoslovakia," Paper 248 read to the ICG conference, Brussels, 1965, p. 2; see also the same writer's "Comments on Archaeological Finds in the Field of Glassmaking in Czechoslovakia," *Czechoslovak Glass Review*, 12, 1966, p. 353. A view of one of the furnaces is given by Karel Hetteš, "Glassmaking in Slovakia–Part III," *ibid.*, 10, 1974, p. 7. A parallel has been drawn with the somewhat later Polish furnace at Kruszwica (see Gasparetto, "A Proposito . . . ," p. 65; and bibliography in JIDV *Bulletin*, 2, 1963, pp. 76–78).

35. Hejdová, "Glassmaking . . . ," pp. 2–3, Figs. 1–3; "Comments . . . ," p. 356; and "The Discovery of a Medieval Glass Furnace in North-Eastern Bohemia," *Czechoslovak Glass Review*, 1, 1957, ill.

36. Hejdová, "Glassmaking . . . ," p. 3; "Comments . . . ," p. 356. Several furnaces, in one instance at least dated to the thirteenth century by pottery, have been excavated at Grünenplan, in Kreis Holzminden, Germany. Of these, three, although of varying dimensions (the longest two meters long and 60 cm wide) were rectangular on plan (W. Haberey, "Glasgeschichte in Scherben," *Schott, Werkzeitschrift*, 5, 1969, pp. 22–23).

always oval, however, is suggested by yet further excavations in Czechoslovakia, this time at Počatky (in the district of Pelhrimov). Here three walls of a furnace were preserved to a height of 40 cm and revealed a rectangular ground plan. This glassworks dated from the late fifteenth or early sixteenth century.[37] A further furnace complex dating from some time later in the sixteenth century has been excavated at Rejdice. Here three rectangular furnaces were found, and the same pattern was confirmed by finds at the nearby contemporary glassworks at Syriste (founded 1558).[38]

Whether the variation between oval and rectangular furnaces in Bohemia is a question of date or of function, it is difficult in the present state of our knowledge to say. In England, the rectangular furnace appears to have been the rule from at latest the fourteenth century onward. At Blunden's Wood a roughly rectangular furnace dating from about 1330 had a central fire trench roughly 10′6″ long by probably originally 2′ wide, with a hearth at either end; to each side of this was a "siege" for two pots, 8′ long by 2′3″ wide and 2′ high. This furnace showed the unusual feature of (apparently) a cavity between the siege and the outside wall, either for purposes of insulation or possibly as a means to the construction of the vaulted roof. This main furnace was accompanied by two subsidiary structures, presumably for the operations of "fritting," "pot arching," and annealing.[39] This picture is repeated in essence at a late sixteenth- to early seventeenth-century glasshouse site at Blore Park, in Eccleshall in northwest Staffordshire. Here a fairly well-preserved stone furnace foundation was found and excavated, revealing a more or less square furnace with a long fire trench running east and west, and a "siege" on either side a foot high, 16½″ wide, and 2′10″ in length, accommodating two pots on each side. This furnace appeared to be complete in itself and no other structure was excavated, although traces were found in neighboring mounds. There is therefore no means of knowing how the ancillary processes were carried out at this site.[40] At the more or less contemporary site of Vann, near Chiddingfold in Sussex, however, the late S.E. Winbolt laid bare the foundations of a larger, brick-built structure, the main furnace being an oblong 12′ by 5′6″, at the corners of which were four diagonally projecting, fan-shaped wings. These were no doubt originally used for annealing,

---

37. Hejdová, "Glassmaking . . .," p. 3; "Comments . . .," p. 356.

38. Hejdová, "Glassmaking . . .," pp. 4–5, Figs. 5–6; "Comments . . .," p. 357.

39. E.S. Wood, "A Medieval Glasshouse at Blunden's Wood, Hambledon, Surrey," *Surrey Archaeological Collections*, 62, 1965, p. 54 ff. See also D. Ashurst and Eric S. Wood, "Glasshouses at Gawber and Blunden's Wood: A Further Note," *Post-Medieval Archaeology*, 7, 1973, pp. 92–94, especially p. 94, where an alternative reconstruction is proposed. A similar furnace, brick-built and almost 21′ long and 2′ 5″ wide, with clearly marked firing chambers at either end, was found well-preserved at Fernfold, Sussex, on a site connected with Jean Carré, founder of the "modern" glass industry in England in 1567; see S.E. Winbolt, "Jean Carré's Glass Furnaces," *Transactions of the Society of Glass Technology*, 20, 1936, pp. 16–18; G.H. Kenyon, *The Glass Industry of the Weald*, Leicester, 1967, pp. 185–188. A comparable furnace of early sixteenth-century date, 12′ long, built of brick and stone and having sieges for three pots each side, was discovered at Bagot's Park, near Abbots Bromley in Staffordshire, and was published by David W. Crossley, "Glassmaking in Bagot's Park, Staffs . . . in the 16th century," *Post-Medieval Archaeology*, 1, 1967, p. 44 ff. One ancillary furnace was excavated in the vicinity. A rectangular furnace site 12′ long by 4′ wide was found at St. Weonard's, in Herefordshire, in 1961. It was probably built of brick and stone, a square of large stones in the center of a burned area probably representing the foundations of the founding-chamber—see N.P. Bridgewater, "Glasshouse Farm, St. Weonard's: A Small Glassworking Site," *Transactions of the Woolhope Naturalists' Field Club*, 37, 1963, pp. 300–315. Further rectangular furnaces, dating from the mid-sixteenth century, with three pots per siege, were excavated at Knighton's, Alfold, Surrey, in 1973 (see *Post-Medieval Archaeology*, 8, 1974, pp. 132–134). Associated with these furnaces was a pair of smaller rectangular furnaces, perhaps for spreading window glass.

40. T. Pape, "An Elizabethan Glass Furnace," *The Connoisseur*, 92, Sept. 1933, p. 172 ff., Figs. I, II, IV.

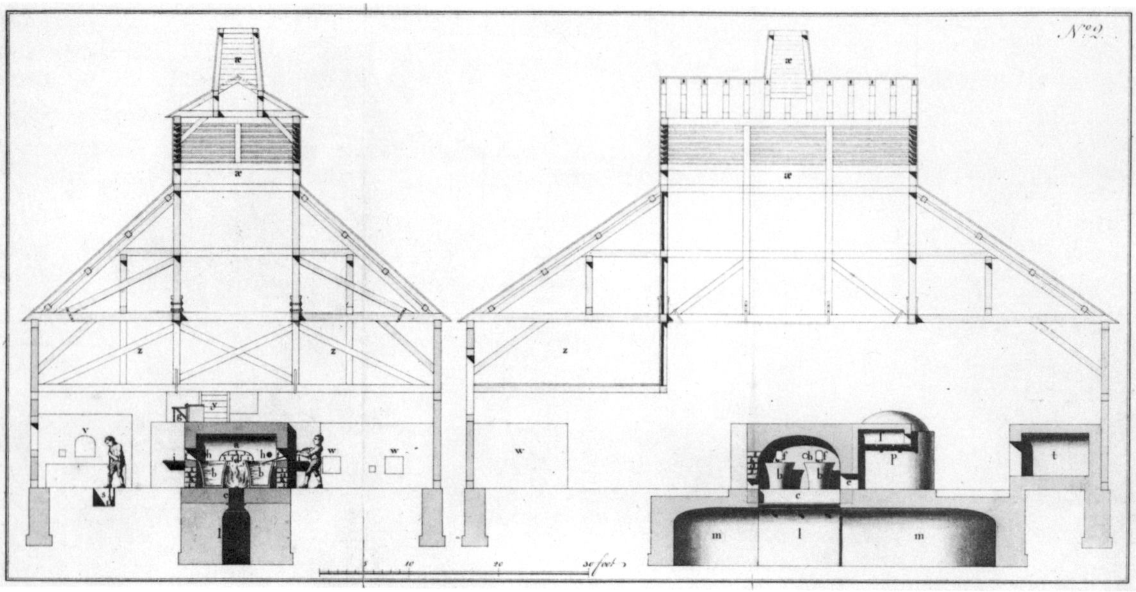

FIG. 17. *View of a bottlehouse in Gravel Lane, Southwark, London, about 1777. Drawing by C. W. Carlberg;* c *represents the grill. See note 42.*

"fritting," "pot arching," etc.[41] All these furnaces made green "forest glass," and their structure is illuminated for us by a passage in Christopher Merrett's translation of Antonio Neri's *L'Arte Vetraria*, published in London in 1662 as *The Art of Glass*. "The *Green Glass furnaces* are made square," he writes on p. 243, ". . . having at each angle an arch to anneal their Glasses. . . .

"For green Glass on two opposite sides they work their Metall, and on the two other sides they have their *Calcars*, into which *linnet holes* are made for the fire to come from the *furnace*, to bake and prepare their *Frit*, and also for the discharge of the smoak. But they make fires in the arches, to anneal their vessels, so that they make all their process in one furnace onely."

This general layout of furnace survived in both England and France until at least the 1770's,[42] although the adapted Venetian-type furnace was by now well-established in both countries. Its characteristic of heating subsidiary furnaces through "linnet holes" takes us back afresh to the two-chamber furnaces of Theophilus and the Mandeville illumination

---

41. Winbolt, *Wealden Glass*, pp. 29–31, with ill.; Kenyon, pp. 193–200, Figs. 17–18, pls. XXI–XXII. The only excavated English furnace which does not fall into the general pattern of a rectangular ground plan, with or without wings, is that at Woodchester, excavated in or before 1904 by the late Basil Marmont (J. Stuart Daniels, *The Woodchester Glass House*, Gloucester, 1950, p. 14). The circular structure shown on the only available plan (*ibid.*, pl. XIV) was apparently 16′ in diameter with an internal diameter of 10′, making the walls 3′ thick, a feature which must arouse some doubt. The firing-hole was 3′6″ wide on the outside and 3′ on the inside, exceptionally wide. A re-excavation, if anything survives, would be highly desirable. An adjacent rectangular structure 9′ by 7′ was interpreted as an annealing furnace (*ibid.*, pp. 3–4). It has been suggested that the round furnace at Woodchester may be explained by the presence there of Flemish glassmakers (W. A. Thorpe, *English Glass*, 3rd ed., London, 1961, pp. 89–91).

42. Diderot and d'Alembert, *Encyclopédie . . . Recueil de Planches . . .*, X, Paris, 1772, pls. 3–6 of article "Verrerie en Bois, Grande Verrerie en Plats"; pls. 8–9 of article "Verrerie en Bouteilles . . . de la Verrerie R[ie] de Seve . . ." (both six-pot furnaces); a four-pot bottle furnace with two (circular) wings was in operation from at latest 1777 in Gravel Lane, Southwark, London (Figs. 17, 18 here) first published by A. Baeckström *et al.*, "Ett Glasbruk mitt i London på 1770-talet," *Glasteknisk Tidskrift*, Årg. 2, Nr. 5, 1947, pp. 76–80, 88–89). Here one wing was used for fritting, the other for heating bottle cullet before its trans-

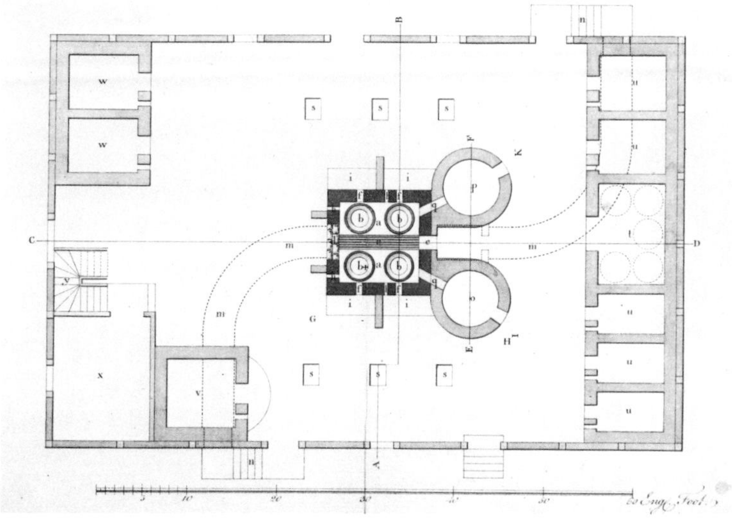

FIG. 18. *Plan of the glasshouse represented in Fig. 17.* qq *are explained as "openings into the vault of these two furnaces, through which the heat is communicated from the founding-furnace":* uu *are annealing furnaces, the largest accommodating 800 bottles;* t *is a pot-arching furnace:* v *is a furnace for chemical apparatus and other large pieces, the annealing furnace for which is at* ww. *See note 42.*

ference to the pots. Annealing was carried out in seven independent subsidiary furnaces (*uu* and *ww* on the plan), and pot arching in another (*tt* on plan). The origins of this type of furnace are uncertain. A furnace with a ground plan of this type excavated at Heindert (Canton d'Arlon, Luxembourg) is supposed to go back "at least" to Carolingian times (C. Dubois, "Un établissement de verriers dans la Forêt d'Anlier," *Annales de l'Institut archéologique du Luxembourg* (Arlon), 46, 1911, pp. 355–373). In England, apart from Vann (see above, p. 23), two winged furnaces have been excavated in recent years: (a) Hutton, in Yorkshire, where an earlier furnace of plain fire trench type was overlaid by a second furnace with two fan-shaped wings to northeast and southwest, these being incorporated with a third overlying two-pot furnace (Fig. 19 here). A second furnace, standing apart, may have been an annealing furnace for the first phase. The last phase of the furnace was dated to the late sixteenth century by thermoremanent magnetism; (b) Rosedale, Yorkshire, where a two-pot furnace of similar date had four fan-shaped wings (Fig. 20 here). Both furnaces were built of stones and clay (see D. W. Crossley and F. A. Aberg, "Sixteenth-century Glass-Making in Yorkshire . . . Excavations at Furnaces at Hutton and Rosedale, North Riding, 1968–71," *Post-Medieval Archaeology*, 6, 1972, p. 107 ff.). From a previous era of archaeology there is record of a furnace excavated at Buckholt (not far from Salisbury) which appears to have had "wings"—E. Kell, in *Journal of British Archaeology*, 17, 1861, p. 55; and E. Wyndham Hume, "English Glass-making in the 16th and 17th centuries, Part IV. From Bowes to Mansell," *The Antiquary*, 31, 1895, pp. 102–106; and A. Hartshorne, *Old English Glasses*, London, 1897, p. 171. The Buckholt glassmakers seem to have been, with one possible exception, French by origin, and the Vann glasshouse appears to have produced green glasses of the normal Wealden "Lorraine" type (thus making improbable the attribution to Ogniabene Lutheri—see Kenyon, pp. 199–200; Eleanor S. Godfrey, *The Development of English Glassmaking 1560–1640*, Oxford, 1975, pp. 32–33, 141). The two Yorkshire sites also produced characteristic green vessels of the same general character. It seems on the whole likely that the furnace with wings was of French origin. That it enjoyed a wider diffusion at a later date is shown by the English examples cited at the beginning of this note, and by the following: (1) a stone-built furnace in a glasshouse at Karlova Hut, in Czechoslovakia, founded in 1758 (Jaroslav Kaván, "Research of Karlova Hut . . . from the 18th century in the Jizera Mountains," *Glass Review*, 3, 1976, p. 10 ff., Figs. 1, 3–4); (2) furnace in the Amelung factory at New Bremen, Maryland, U.S.A., built after 1774 and probably after 1784 (Ivor Noël Hume, "Archaeological Excavations on the Site of John Frederick Amelung's New Bremen Glassmanufactory, 1962–1963," *JGS*, 18, 1976, p. 138 ff., especially p. 150, Fig. 3 and pp. 159–161). Both these furnaces appear to have had a plain firing trench, without a grill (see p. 26 below), no doubt reflecting the use of wood rather than coal as fuel. Amelung had begun his career at Grünenplan, in Germany, south of Hanover, and in Maryland he took over a factory that had previously been run by Germans. Therefore we may reasonably conclude that the winged furnace was established also in the German-speaking countries by at latest the middle of the eighteenth century. It seems likely that a six-pot furnace with two annexed annealing furnaces was used at the Notsjö factory in Finland in 1799 (Jacob Seela, "The Early Finnish Glass Industry," *JGS*, 16, 1974, p. 65, Fig. 3).

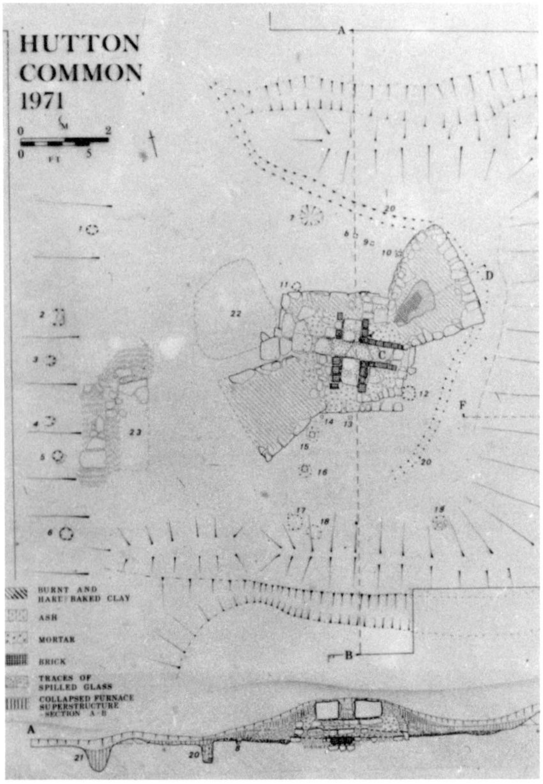

F IG. 19. *Plan of a green-glass furnace excavated at Hutton Common, Yorks (after D. W. Crossley and F. Aberg). See note 42.*

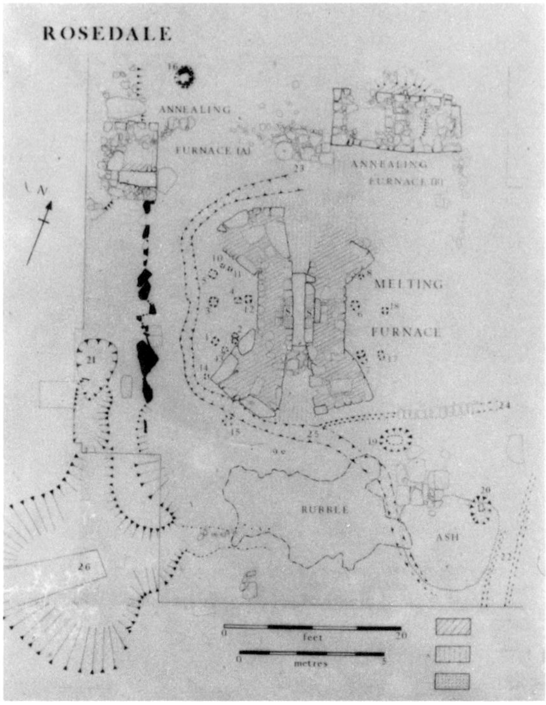

F IG. 20. *Plan of a green-glass furnace at Rosedale, Yorks (after D. W. Crossley and F. Aberg). See note 42.*

(Figs. 15, 16). This plan appears to have survived in Germany until at least the end of the seventeenth century.

Amongst the illustrations to Johann Kunckel's *Ars Vitraria Experimentalis*, first published in Amsterdam and Danzig in 1679, appear engravings showing the German furnace (Figs. 21, 22). Here the working furnace and the fritting or annealing furnace are simply two chambers of the same structure, with the fire chamber running throughout their combined length (Fig. 22). It is not clear from the engravings whether or not there was also a "linnet-hole" connecting the upper chambers (that is, the working "story" of the main furnace and the actual annealing chamber of the subsidiary one), and Kunckel provided no explanatory text. Fortunately, these illustrations were repeated in the 1752 French edition of Neri, and were there explained.[43] That the type of furnace represented was continuously in use up to the date of Kunckel's book is indicated by the results of an excavation carried out on the site of a glasshouse at Trestenshult, in Sweden, dating from about 1630. That it was essentially a German-type furnace is confirmed by the fact that this factory was under the leadership of a certain Påvel Gaukunkel, probably one of a West German family of glassmakers, and that it made green glass of German "Waldglas" type. This furnace had an octagonal founding furnace with four pots, and a stoke-hole running throughout the length of the whole structure, which included an annealing furnace of circular plan, set slightly askew to the axis of the main furnace, with the result that the square hole permitting the passage of heat from the fire chamber to annealing

FIG. 21. *View of Kunckel's furnace (from the 1752 French translation of A. Neri,* L'Arte Vetraria*).*

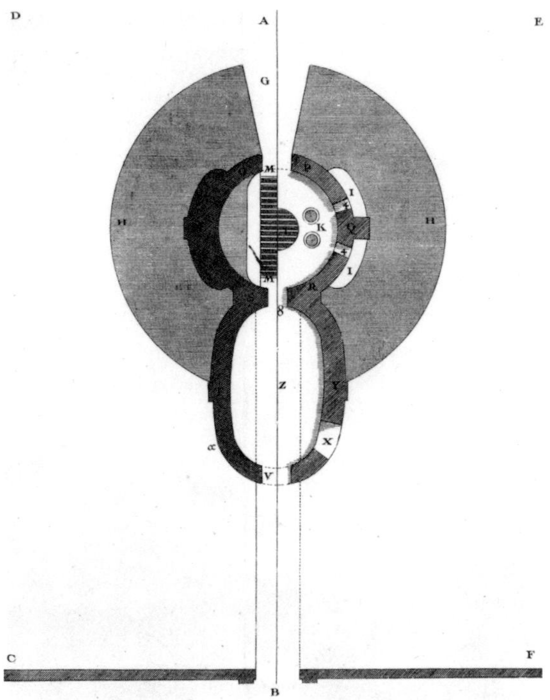

FIG. 22. *Plan of the furnace shown in Fig. 21.*

chamber is off-center (Fig. 23). The two furnaces are interconnected at "first floor level" by a "linnet-hole" (K of the drawing, Fig. 23).[44] A furnace of the same basic type was known to Agricola already in the sixteenth century, although here the annealing furnace is rectangular in plan—perhaps a throwback to the earlier traditions of Eraclius and Theophilus (Fig. 24). Agricola describes this in the following terms:

> The *second furnace* is rounded, ten feet broad and eight high, and strengthened on the outside with five ribs 1½ feet thick. This again consists of two chambers, the roof of the lower being 1½ feet thick. This lower chamber has in front a narrow opening for stoking the logs on the ground-level hearth; and in the middle of its roof is a big

---

43. *Op. cit.* in n. 23, p. 618: "STVXYRZ. Plan du fourneau à calciner ou à recuire; il est contigu au précédent (i.e., the founding furnace) avec lequel il communiqué par l'ouverture 8; cette ouverture est au-dessus du plan où posent les creusets, ensorte que la flamme va du premier fourneau dans le second." It is important to note that the subsidiary furnace was for annealing as well as for fritting; the opening marked "X" is glossed as "ouverture pour placer le verre à recuire." This type of furnace was clearly domesticated in France about as early as the "winged" furnace was established in Germany/Bohemia, for among the Supplementary Plates of Diderot's *Encyclopédie* is a series of engravings illustrating this type of double furnace, entitled interestingly enough "Four rond ou à la Française." There are changes of detail, notably a square plan for the subsidiary furnace (see at left), and four ribs to the beehive founding furnace (cf. Fig. 21).

44. H. Seitz, "Faktorier och Manufakturer . . . ," in *Svenska Folket Genom Tiderna,* Malmö (c. 1939), pp. 279–282. Gaukunkel, although referred to as a Dutchman, seems much more likely to have been of German origin, the termination "-kunkel" being highly suggestive (H. Seitz, *Glaset Förr och Nu,* Stockholm, 1933, p. 94). In 1631 Gaukunkel was summoned to Russia, and arrived on the island of Dagö (modern Hiiumaa in Estonia) to take over an establishment previously run by Jost Wentzell, of Hessian origin. Gaukunkel's furnace at Hüti appears to have repeated the general pattern of that at Trestenshult, except that the subsidiary furnace was long and narrow (Maks Roosma, *Hüti Klaasikoda Hiiumaal,* Tallinn, 1966, Fig. 20; see also *idem,* "The Glass Industry of Estonia in the 18th and 19th century," *JGS,* 11, 1969, pp. 70–72; also *JGS,* 19, 1977, pp. 183–184). The chimney at the far end of the reconstruction is probably a misinterpretation of the evidence of the masonry. A second furnace on the site was probably a fritting furnace (Roosma, *Hüti . . . ,* Fig. 28).

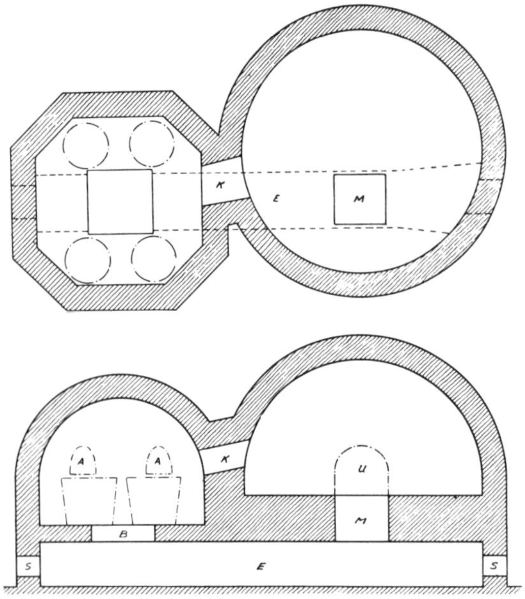

FIG. 23. *Reconstruction of a green-glass furnace found at Trestenshult, Sweden, about 1630 (after E. Strömberg).*

FIG. 24. *Woodcut illustrating G. Agricola,* De Re Metallica, *Basel, 1556, showing his "second furnace" with conjoined annealing furnace.* H *shows the clay "tunnels" in which the glasses were to be put to be annealed.*

round aperture opening into the upper compartment so that the flames may penetrate into it. But in the wall of the upper chamber between the ribs there should be eight windows so large that through them the bellied pots may be put on the floor round the big aperture. . . . At the back of the furnace is a square opening, in height and breadth 1 palm, through which the heat may penetrate into the *third furnace*, adjoining. This is oblong, eight feet by six broad, similarly consisting of two chambers, of which the lower has an opening in front for stoking the hearth. . . .[45]

This is in all essentials the Trestenshult furnace except for the ground plan and for the fact that apparently each furnace here has its own firebox, one set at right angles to the other. But the express mention of interconnection at "first floor level" is the vital point here, and we may suspect that it existed also in the furnace of Mandeville's illustration (Fig. 16).[46]

---

45. Winbolt, *Wealden Glass*, p. 80.

46. Evidence for further furnaces of this type is to be found in the German-speaking areas of Central Europe: (1) the well-known Preussler *Humpen* in the Museum of Applied Arts, Prague (Figs. 25, 26 here), dated 1680 (see A. von Saldern, *German Enameled Glass*, Corning, 1965, p. 137 and Fig. 234); (2) a second Preussler family *Humpen* of 1727, shows a virtually identical furnace (Fig. 27 here). To the right may be seen the subsidiary furnace with something stretched across the access door, as in the case of the 1680 *Humpen*, which seems to show pot arching in progress. To the left is a separate secondary furnace (for fritting?); (3) a glasshouse at Reichenau (probably that on the Buquoy estates at Gratzen on the Austrian side of the south Bohemian border) shown in Fig. 28 here. The house clearly made both vessel and window glass (the latter by the muff process), the circular furnace for vessel, the structure to its left being designated on the key as the annealing furnace, and that to the extreme left as the "Taffel Offen," presumably the furnace for window glass. The structural connection, if any, between these two last is unclear from the engraving, but the lehr is clearly connected with the circular vessel-glass furnace, although it seems to be in use for annealing (and perhaps spreading?) muffs. Probably this should be regarded as a bastardized version of the "German" furnace. The engraving by Clemens Beuttler was used for the illustration of M. Merian, *Beschreibung der Herrschaften . . . Richenau*, Frankfort, 1656 (I am indebted to Dr. Olga Drahotová for the photographs of the Reichenau furnace and for advice about it); (4) the illustration from the Augsburg Bible of 1730, showing all the features of the Kunckel engraving, and perhaps derived directly from it (ill., e.g., *Glass from the Corning Museum of Glass: A Guide to the Collections*, Corning, 1955, frontispiece); (5) excavations of the Junkernfelde furnace (see below).

FIG. 25. Humpen *of Christian Preussler with view of the Zeilberg glasshouse, Bohemia, 1680. Museum of Applied Arts, Prague. See note 46.*

FIG. 26. *Another view of the* Humpen *shown in Fig. 25. See note 46.*

FIG. 27. Humpen *of the Preussler family; 1727 (after Partsch, "Der Preusslerhumpen," 1928). See note 46.*

FIG. 28. *Engraving showing glasshouse at Reichenau, from a drawing by Clemens Beuttler. Middle of seventeenth century. See note 46.*

III. Coal Firing

All the furnaces hitherto described were fired by wood. It fell to England in the sixteenth century, with the onset of rapid deforestation and the consequent dread of depriving the Navy of timber, to develop the use of coal for this purpose. The late sixteenth and early seventeenth centuries were filled with the clamor of rival patentees claiming monopolies for their particular systems of using coal for this and that industrial process. In glassmaking sufficient progress had been made by 1615 to enable the government of the day to issue a "Proclamation touching Glass," forbidding the burning of wood in the glass industry.[47] The essential feature of the coalburning furnace was the use of a grill of iron bars on which the coals could be laid and raked periodically, the ashes falling into a pit below. Merrett, writing in 1662, says (p. 364): "Sleepers are the great Iron bars crossing smaller ones which hinder the passing of the coals, but give passage to the descent of the ashes." This appears to be the earliest reference to this device, but it is clearly shown in the drawing of a London glasshouse made by a Swedish architect visiting England in 1777–78 (Fig. 18). The grill system seems to have been adopted on the Continent without much delay. It is shown in the illustration to Kunckel's *Ars Vitraria Experimentalis* of 1679 (Fig. 22) and is clearly to be seen in the illustrations to the great French *Encyclopédie*. It will be observed that the provision of an ash chamber added, as it were, an extra story to the furnace.[48]

Clearly the provision of an adequate draft was of critical importance in the firing of a coal furnace, and it was inevitably in this field that the English made a further important contribution to glass technology. The principle was to have a cone-shaped building to house the furnace, capable of having all its outlets closed. Underground flues, no doubt arranged with regard to the prevalent wind, led to the furnace itself. By closing the cone and opening

---

Three sites excavated in Denmark seem to suggest that this "German" type of furnace was also in use there. (1) Hyttekaer, in Tem Glarbo, in the general area of Aarhus, Jutland. Here six furnaces were discovered, not all necessarily of the same period, of which two appeared to have a common fire trench, one being a circular, the other an oblong structure. Two further furnaces were apparently circular on plan, but although close to each other, their fire trenches were aligned roughly at right angles. Of the remaining furnaces apparently only the fire trenches survived, without any indication of their external shape (Thelma Jexlev, Peter Riismøller and Mogens Schlüter, *Dansk Glas i Renaessancetid*, Copenhagen, 1970, pp. 83–86, 148–152; pls. 1, 1; 2, 1; 3, 1; Figs. 11, 3; 37–39); (2) Tinsholt, south of Aalborg, in Jutland. At least one furnace (or two contiguous furnaces) here apparently corresponds to the main Hyttekaer furnace, although the published details are somewhat unclear (*ibid.*, pp. 80–83, 152–156; pl. 1, 2; Fig. 40); (3) Stenhule, in Tem Glarbo (see [1] above). Three furnaces were found (discounting an apparently earlier circular foundation) of which the founding furnace, probably for six pots, may have been almost square (11′ x 9′), with chamfered corners. Its very long firebox, projecting on the west side, may have heated a long, narrow subsidiary (annealing?) furnace (*loc. cit.*, Fig. 44), which would have resembled that at Reichenau. The two remaining furnaces, apparently circular and oval on plan respectively, were possibly for fritting and spreading of window glass (*ibid.*, pp. 87–93, 158–161: pl. 1, 3; Figs. 3, 2; 10, 3; 11, 3; 14, 1; 43–44). A warning note should be struck here. The site at Nejsum, in Vendsyssel, north of Aalborg, the plan of which seems clearer than some of those mentioned, reveals a central fire trench projecting a long way out at either side of an almost circular furnace plan—very much as at Fernfold, in the Weald, except that the furnace there was rectangular (see above, n. 39). The Stenhule furnace could well be reconstructed in the same sense (Jexlev, *op. cit.*, pp. 90, 160). It should be noted that all the glassmakers associated with these Jutland sites were Germans, some specifically from Hesse (*ibid.*, pp. 33, 43, 46, 53, 58–59, 64–66).

A glasshouse at Henrikstorp in Skåne, Sweden, had a ground-plan suggesting an exceptionally long fire trench, and may have been a structure resembling that at Reichenau. Its reconstruction with four pots on one side of an offset fire channel, and with a circular annealing furnace, seems very improbable (S. E. Noreen and H. Graebe, *Henrikstorp: det Skånska Glasbruket, 1691–1760*, Göteborg, 1964, p. 53 ff.).

Finally, a glasshouse dating from about 1530 excavated at Junkersfelde, in eastern Westphalia, revealed a four-pot furnace rectangular in plan, with chamfered angles (as at Trestenhult and Stenhule), apparently leading to an annexed annealing furnace (Ursula Wichert-Pollmann, *Das Glasmacherhandwerk im östlichen Westfalen*, Münster, 1963; Jexlev, *op. cit.*, pp. 139–140).

47. For a discussion of events in this period, see Godfrey, *op. cit.*, p. 38 ff., esp. pp. 59–74. Unfortunately, no de-

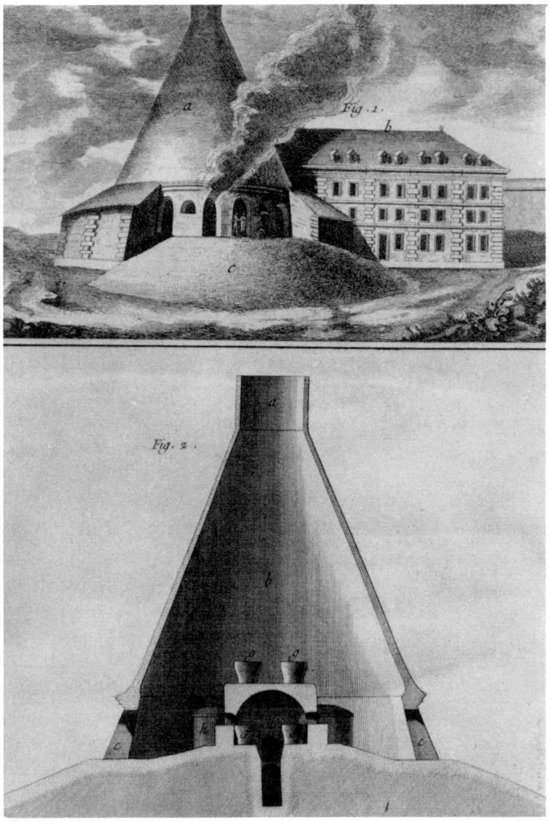

FIG. 29. *Plate from Diderot's* Encyclopédie *showing "Verrerie Angloise, Vue et Coupe sur la Longueur de la Verrerie" (pl. vol. X, pl. 2 of article "Verrerie Angloise").*

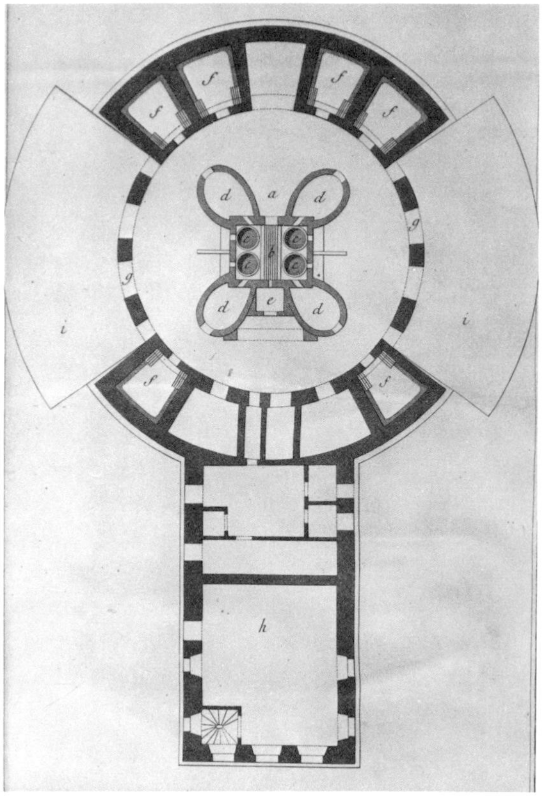

FIG. 30. *Plan of the glasshouse shown in Fig. 29 (*ibid., *pl. 1).* b *in Fig. 2 is the grill. The "linnet-holes" to the "wing" furnaces are visible in the engraving.*

the flues, a tremendous through-draft was created, the great cone acting as a chimney (Figs. 29–30). The advantages of this system are described in the French *Encyclopédie*. The writer of the article *Verrerie* stressed the importance of the single upward movement of air facilitated by the cone superstructure, all the outward exits of which were shut during founding, allowing the air to enter the building only through three broad ducts laid in the foundations under the furnace. He calculated that this increased the efficiency of the English furnace enormously in comparison with that of the French type, the former being able to make in twelve days, other things being equal, the same number of bottles as a French factory made in fifteen. The same points are chosen for comment by a Danish visitor to London in 1727: "It is known, that the English glass is of high quality, and this is ascribed to the intense heat of the coal which is used for this purpose; I therefore also visited two of their glass-furnaces, which stand on an eminence, and in the ground there is an open passage from both sides to a very large iron grill, which is in the middle, and the coals on and about it; by means of this passage the coals receive the air they need and one can stir them with appropriate

scription of Percival's patent processes appears to have been recorded.

48. *Recueil de Planches* . . ., 10, Paris, 1772, pls. 2–3 of article "Verrerie en Bois" (*sic.*—the principle had by now been extended to wood-burning furnaces), and pl. 1 of "Verrerie Angloise" (Fig. 29 here).

instruments.[49] It is not certain how long before this the glass cone was devised. Captain Philip Roche was building one in Dublin as early as 1696,[50] and in 1702 *The London Gazette* reports one "94 Foot high and 60 Foot broad."[51] It seems probable that a period of experiment preceded these efforts. Possibly the great heat reported by Merrett (p. 17 above) was due to this device even as early as 1662.[52] More probably, the cone was invented between this date and the end of the century. A few still survive, though no longer used.[53]

One further English invention resulted from the use of coal for firing. This was the covered glass pot. That open pots continued to be used in circumstances where the fumes of the coal made little difference (e.g., in bottle making) is proved by the details of the interiors of English bottle houses given by C. W. Carlberg in 1777–78 (Fig. 17) and by the illustrator of the French *Encyclopédie* in 1772.[54] Here the old type of open pot is clearly visible. It seems more likely, therefore, that the development of the covered pot was associated with the evolution of lead crystal, for this is irreparably damaged by the sulphur compounds which result from the combustion of coal.[55] Curiously enough, the earliest mention of it appears to come from Norway. In May, 1756, at the Nøstetangen factory, there were prepared "16 English covered pots."[56] The earliest pictorial representation which the present writer has been able to discover dates from as late as 1802.[57]

POSTSCRIPT

The original thesis of this article—that there was a "Southern" tradition of furnace building in which the founding-furnace was normally circular and incorporated a third "story" for annealing, and a "Northern" tradition in which the founding-furnace was normally rectangular or oblong on plan and the annealing-furnace

---

49. *Claus Seidlins Optegnelser om Sit Levned, 1702–82*, Copenhagen, 1915, kindly communicated by Dr. Ada Polak. Godfrey, *op. cit.*, p. 147 ff. has an interesting discussion of this development, including some eyewitness descriptions of the Winchester House furnace, probably from as early as about 1610. There is, however, some misunderstanding of the functions of the glass cone, which is quite different from a chimney leading from the furnace. The great point is that the heat shall be drawn over the pots and not out of the crown of the furnace. The cone and the underground passages simply augment the draft, which follows this course, emerging at the working holes of the furnace. An early example of a four-pot furnace with an ash pit below the sieges, and an apparently vaulted firing chamber, was excavated by Ruth Hurst Vose and Freda J. Burke for the Pilkington Glass Museum at Haughton Green, Denton, near Manchester. References to glassmakers in local parish registers date within the first half of the seventeenth century (*JGS*, 12, 1970, pp. 184–185; 13, 1971, p. 149; *Post-Medieval Archaeology*, 4, 1970, pp. 185–186).

50. M. S. Dudley Westropp, *Irish Glass*, London, n.d., 1920, pp. 37–38. The thirty-five foot building constructed in 1621 at Ballynegery and mentioned by Mrs. Godfrey, *op. cit.*, p. 156, was evidently a normal frame-and-shingle glasshouse with presumably no effect on the draft (Westropp, p. 30). The *Belfast News-Letter* for August 19, 1785, referred to a new glasshouse 120′ high "being the largest of any in Great Britain or Ireland." In 1823 a cone 150′ high was recorded in the same city (D. Westropp, *Irish Glass*, pp. 106, 109). In 1784, a prohibition was promulgated in Dublin against any glasshouse chimney less than 50′ in height (*ibid.*, p. 47).

51. See D. R. Guttery, *From Broad-Glass to Cut Crystal*, London, 1956, p. 38.

52. Merrett, *op. cit.*, p. 245. The difference between the heat needed for making drinking glasses and that for window glass is made explicit in an entry in the diary of Sir James Hope in 1647 concerning the Wemyss (Scotland) glasshouse: "That window glasses and drinking glasses can not be made in one fornace because those requyre a great deall stronger heatt than these; That the fornace for those is y^r^fore long vaulted; and for these round bot however y^t^ he could not make window glass, nather possiblie could find workemen who have skill of both. . . ." (*cit.*, R. Oddy, "Scottish Glass," *Glass Circle Papers*, No. 151). These observations were based on the experience of an Italian glassmaker named Christopher Visitella. The great heat reached in the green-glass furnaces is emphasized by Merrett: "The heat of those *furnaces*, is the greatest that ever I felt. . . . The workmen say 'tis twice as strong as that in the other Glass-furnaces . . ." (*op. cit.*, p. 245).

53. For example, at Catcliffe, in Yorkshire (Geoffrey D. Lewis, *The South Yorkshire Glass Industry*, Sheffield, 1964, pl. 1). The foundations of a glasshouse of this type have been excavated at Gawber, also in Yorkshire. It was probably constructed in the eighteenth century and was "in ruins" in 1823. It illustrates in practice the use of underground

often on a level with it and interconnecting by means of a common fire-channel, and usually also by "linnet-holes"—seems in general to have been borne out by subsequent discoveries. The "Northern" tradition clearly bifurcated at some point, probably in the sixteenth century, into those furnaces which were built with wings for subsidiary firing processes on the one hand and, on the other, two-chamber furnaces of the type represented by the Mandeville manuscript and J. Kunckel's engravings. These general developments have to be viewed in the context of all the variables imposed in particular cases by site, available building materials, etc. It is to be hoped that in all future excavations on glasshouse sites, proper measured drawings accompanied by photographs will illustrate the report, together with details of the refractories used (pots, *bocca* frames, etc.) and of the glass items found on the site. All these are essential clues to the nature of the furnaces used.

---

flues to induce the draft under the furnace (Denis Ashurst, "Excavations at Gawber Glasshouse, near Barnsley, Yorkshire," *Post-Medieval Archaeology*, 4, 1970, pp. 92–140). Here, on the site of the main furnace, were found "lengths of fire bar, 2 to 3 ft. long by 1¼ in. square section and a sandstone block carved with a slot into which a fire bar exactly fitted, to form the detachable fire grate . . ." (*ibid.*, p. 106). This seems to be the only archaeological record of the dimensions of these bars.

54. See Figs. 17, 18, 29 here.

55. Discussed by Mrs. Godfrey, *op. cit.*, p. 148 ff.

56. Ada Buch Polak, *Gammelt Norsk Glass*, Oslo, 1953, p. 247, n. 36.

57. *Versuch einer ausführlichen Anleitung zur Glasmacherkunst*, Frankfurt, 1802, pl. II.

The Houghton Box. Gold, engraved, and with translucent enameling; containing two cut-glass flacons. Probably German, late 17th or early 18th century. H. 4 cm, L. 4 cm, W. 3.2 cm. Collection of Mr. Arthur A. Houghton, Jr.

II. A GOLD AND ENAMEL BOX IN THE FORM OF A GLASS FURNACE

Robert J. Charleston

THE GOLD BOX which forms the subject of this article was acquired by Mr. Arthur A. Houghton, Jr. in 1970[1] and appears to be unique. Its practical function is to house two faceted crystal glass scent- or smelling-bottles and their filling-funnel, but its outward appearance is much less prosaic. Here, in miniature (on a scale which indeed makes the scent-flasks seem grotesquely large when seen in juxtaposition) is represented no less than the whole life of a glasshouse.

The box is 4 cm long, of the same height, and 3.2 cm deep from back to front. On plan a rectangle with chamfered corners, it has a domed cover which opens on a nine-part hinge and fits over a slightly offset flange on the upper part of the body of the box. It is secured by a rough iron catch activated by a button in the front, and (later?) iron screws to either side presumably anchor the lining of the box (Fig. 1). The base-plate of the box projects slightly beyond the vertical sides, and on this tiny ledge are perched with great skill the figures of a number of glassworkers, whose activities will be described in a moment. Their costume is picked out in blue, green, and purple translucent enamels. The interior of the box is fitted with a lining, the top of which, pierced with two circular holes to accommodate the bottles, is faced with a now gray napped textile material edged with a gilt cable-gimp. The cavities to receive the bottles are lined with cardboard.

The scent- or smelling-bottles which are the ostensible *raison d'être* of the box, are octagonal on plan, vertical-sided, with wheel-faceted angles and shoulders. The swelling gold caps have a gadrooned upper surface and a collar which fits over the neck of the bottle. The caps have fixed internal pins, one of which is mounted with a tapering cork secured by a tiny nut threaded on the end of the pin; presumably both were originally fitted in this way. To fill the bottles there is a tiny funnel with oval receptacle.

The vertical sides and rim appear to be made of metal strips soldered at the angles and strengthened in places by applied horizontal moldings. The domical cover has apparently been raised from a single plate, with the "chimney" and cowls soldered on. The "chimney" is made from a strip repeating the treatment of the vertical parts of the furnace, but with a second strip fixed over it like a flat handle, and the cowls are made from four curved lenticular pieces of gold (Fig. 2). The whole of the outside of the box and the base have been engraved, the former to represent

1. Parke-Bernet sale, November 20, 1970, Lot 43 (illustrated in the Catalogue).

FIG. 1. *Front view of the Houghton box showing contents.*

FIG. 2. *Top view of the Houghton box showing engraved designs.*

FIG. 3. *Base of the Houghton box, showing engraved designs.*

FIG. 4. *Back view of the Houghton box with workmen pounding and making pots(?).*

brickwork, the latter with a central motif of curved acanthus leaves radiating in a symmetrical arrangement from a central cross-hatched roundel (Fig. 3). Following the edge of the baseplate are two laurel wreaths tied at top and bottom with crossed ribbons, the tips of their leaves meeting in the centers of the two short sides (Fig. 3). The symmetrical laurel leaf motif is repeated down most of the angles of the "masonry" on the outside of the box, creating in appropriate places a pleasing ambiguity by suggesting flames issuing from glory holes, etc.

The box and its contents are struck in a number of places with the following marks:

(1) A *fleur de lis* in a compartment which follows its outlines. This, apparently a quality mark for gold, has never been decisively interpreted.[2] Its appearance, however, on objects of late eighteenth- or early nineteenth-century date[3] guarantees that it is not the mark of the maker of the box, nor the hallmark of a city guild contemporaneous with it.

(2) A capital "A" standing slightly askew in a lozenge-shaped compartment. This is a duty mark for objects of 20 carat gold used in Vienna between 1806 and 1807.[4]

(3) An eagle's head erased, in a roughly oval compartment. This mark is a restricted warranty mark for gold, used in Paris from May, 1838, onward.[5]

(4) A mark probably of a "V" in a shaped panel. The "V" mark appears to be one used in Holland from 1909 onward on objects without a statutory metal content.[6]

As has been suggested above, the box represents all the processes which went on in a glasshouse. It may be considered logical to start at the beginning, which in this case means the back of the box (Fig. 4), where two men may be seen at work, the one on the right busy pounding in a mortar, the one on the left handling a cylindrical object from which he appears to be standing back slightly in order to assess it. The figure on the right wears a green tunic and purple breeches and cap. He raises a heavy pestle in both hands to pound the substance in a mortar standing on a column. Evidently he is reducing to powder either coagulated masses of ash, calcined quartz, or lumps of frit. Peder Månsson, who visited Italy between 1508 and 1524, described how the marine soda weed used as a glass flux was collected and burned in pits and then water thrown over it—"till the pit is full. The ash then lies, black, collected in big lumps, while at the bottom of the pit is the salt (*sal alkali*) like a grey stone. . . . Take the salt stone, dry as you find it, pound it fine in a mortar, and pass it through a fine sieve. . . ."[7] In the Northern countries, where the ash was more likely to be derived from burned land plants or wood, the

---

2. Marc Rosenberg, *Der Goldschmiede Merkzeichen*, Berlin, IV, 1928, p. 201, no. 5829.

3. Serge Grandjean, *et al.*, *Gold Boxes and Miniatures of the 18th Century* (Catalogue of the James A. de Rothschild Collection at Waddesdon Manor), Fribourg, 1975, p. 92, no. 43.

4. Marc Rosenberg, IV, 1928, p. 435, no. 7880.

5. L. Carré, *A Guide to Old French Plate*, London, 1971 reprint, p. 208.

6. Rosenberg, IV, p. 394, no. 7557. The marks are struck on the components of the *étui* as follows:

Caps of the bottles, marks (1), (3) and (4)
Filling-funnel, marks (1), (2) and (4)
Body of box, mark (2) and other indecipherable marks
Cover of box, marks (2) and (4) and other indecipherable marks.

The indecipherable marks, of course, may be any of those recorded above, indistinctly struck.

7. See *op. cit.* in n. 10 of previous article, pp. 77–78. James Howell, who was in Alicante (a favored source of *sal alkali*) in 1621, describes how the blue berries of the *barilla* "melt and dissolve into an Azure Liquor, and fall down into the Pit till it be full; then they dam it up, and some days after they open it, and find this *Barillia* Juice turn'd to a blue Stone, so hard, that it is scarce malleable . . ." (*Epistulae Ho-Elianae* . . ., London, 1737, p. 51). This blue color was recognized as the most desirable at least as early as 1509, in a letter written to a Venetian representative in the Near East: "Quelle buone devono . . . avere un bel colore celeste, cioè azzurrognolo ("biavo"), non nero. . . ." L. Zecchin, "I primi cristalli Muranesi giunti in Oriente," *Vetro e Silicati*, XII, no. 71, 1968, p. 19. Antonio Neri (*L'Arte Vetraria*, Florence, 1612, p. 1) prescribes that the salt-stone be ground in a stone mortar with an iron pestle.

process of lixiviation in an attempt to purify the ashes would probably produce a similar result, calling for the grinding of the flux before it was fritted; indeed, Kunckel describes how a mixture of water and ash is boiled in a metal pan, leaving behind a salt which has to be broken out of the container[8] (and therefore probably requiring to be ground before fritting). In a similar way, the other constituent of the frit—the silica (quartz or sand)—was in the best practice often also ground before fritting. Antonio Neri may be our guide here, prescribing: "The whitest Tarso, which hath not black veins, nor yellowish-like rust in it. At Moran (i.e., Murano) they use the pebles [*sic*] from Tession [Tesino] . . . *Tarso* there is a kind of hard, and most white marble, found in Tuscany. . . ."[9] The "tarso" had to be pounded small and sieved before being mixed with the prepared salt in the ratio 200:130 preparatory to the fritting process. Kunckel, dealing with German practice, says bluntly that sand will do just as well, or flints from various parts of Germany, which should be calcined and then thrown into water "the better to pound them."[10]

FIG. 5. *Interior of a bead factory in Amsterdam(?), showing workmen pounding materials in mortars. By Jacob van Loo. Royal Museum of Fine Arts, Copenhagen.*

When both silica and alkali had been ground up as small as required, sieved, and mixed in the desired proportions, they were laid on the floor of the fritting-furnace ("calcar," "Frittoven," etc.) and subjected to a fierce reverberatory fire. This drove off "some of the gaseous products from the melting together of the batch materials and burns off the carbonaceous impurities in the fluxes."[11] The mixture had to be assiduously turned by a man using a long iron rake. This process is not represented on the box, no doubt because it normally took place in a separate furnace.[12] When the frit had reached a semi-molten pasty consistency, it was raked from the fritting hearth and usually soused with water to purify it (Neri prescribes "three or four buckets").[13] Månsson, however, writing almost a hundred years earlier, in a more primitive phase of the art, says: "When the furnace has become cold, the ashes are taken out, pounded and powdered very fine, and sifted like flour. This is the method of preparing the frit."[14] One way or another, there must have been a great deal of pounding to be done, and the energy consumed in this way may well be seen in G. M. Butteri's painting of the Florence glasshouse in the *studiolo* of Francesco I de' Medici (Fig. 6 in the previous article, center middle-foreground). Similar work in progress, but somewhat more languid in execution, may be seen in Fig. 5 here.

So we turn to the second of the two figures at the back of the box, in his pale blue tunic, bright blue cap, and green breeches. One might be tempted to suppose that he is put where he

8. Joh. Kunckel, *Ars Vitraria Experimentalis*, Frankfurt and Leipzig (1679), pp. 347–348.

9. Christopher Merrett, *The Art of Glass*, London, 1662, p. 7 (translation of Neri, Book I, chap. II).

10. "Um desto besser zu stossen," see Kunckel, *op. cit.*, pp. 47–49.

11. R. W. Douglas and S. Frank, *A History of Glassmaking*, Henley-on-Thames, 1972, p. 101.

12. R. J. Charleston, "Glass," in *A History of Technology*, 3, ed. C. Singer *et al.*, Oxford, 1957, pp. 210–211, Fig. 141.

13. Neri, *op. cit.*, p. 10.

14. *Op. cit.*, p. 78.

is for the sake of symmetry, and this consideration may indeed perhaps dictate the use of a pillar on which he holds the object about which he is busied. He has the air of a potter standing back from his work to catch in proper perspective the form on which he is engaged. There seems to be every likelihood that he is in fact the pot-maker of the establishment. Vannoccio Biringuccio described the pots of his day (1540) as being thrown on the wheel,[15] but this seems unlikely at the presumed date of this box. Christopher Merrett, in his notes on the text of Neri, writes: "This (i.e., Stourbridge) clay being well washed from all impurities is calcin'd in a *furnace* for this purpose, and then ground in their mill into a fine powder, which being mixed with water is trod with their bare feet till it come to a good consistence, fit to mould, which they do with their hands, and when fashioned, dry them in a convenient place, and afterwards anneal them in or over the furnace."[16] The building up of the pots by hand from carefully prepared rolls of clay may be seen in the illustrations to the great *Encyclopédie* of Diderot and d'Alembert[17] (Fig. 6 here). It will be observed that the pot is being shaped on a circular base which may perhaps be considered the equivalent of the pillar represented on the box. The pots of the seventeenth and eighteenth centuries were normally cylindrical or bucket-shaped, and of many different dimensions, and there is no incongruity in either the shape or the size of the pot represented on the box, although the pot rim was not usually profiled in the way shown, being normally turned inward if at all.

The glass material and the pots having been prepared in the ways described, and the pots preheated and finally moved into position in the main furnace, the fire was stoked up to hasten the melting and "fining" of the glass in the pots, and on one short side of the box we see the *tiseur* (in English "teasehole man") crouched down to perform his office of feeding the fire with billets of wood (Fig. 7). He is clad

FIG. 6. *Making pots, from Diderot and d'Alembert,* Encyclopédie, *pl. vol. X, 1772, pl. 8, "Verrerie en bois. Différentes opérations pour la formation d'un Pot de petite Verrerie à Pivette."*

FIG. 7. *Side view of the Houghton box, showing* tiseur.

---

15. Charleston, *op. cit.*, p. 210.

16. *Op. cit.*, pp. 245–246.

17. *Recueil de Planches . . .*, X, Paris, 1772, pl. 8 of article "Verrerie en Bois . . . Petite Verrerie à Pivette." This procedure was still favored in the nineteenth century, and continued in use in the present century (D. Lardner, *Cabinet Cyclopaedia*, London, 1832, pp. 159–160).

in green tunic, pale blue breeches, and purple cap. The billets were precut to the requisite length, matching the width of the fire-channel, and were fed into the fire at a stoke-hole (French *tisard*) which was often sunk below ground level. The wood was frequently stacked on a wooden staging over the furnace itself (see p. 16 of the preceding article), and in the advanced French glasshouses illustrated in the great *Encyclopédie*[18] there was a chute down which the billets were fed to the "teasehole" men below. Feeding the fire was one of the most tedious occupations about the glasshouse, and those who undertook it either tended to be a little mad, or became so.[19] With furnaces like the German model, stoked from both ends, the stoker had to make a ceaseless round from one end to the other and back again. There was normally one man for the day shift and one for the night, and a Swedish *tiseur* is recorded who could run his endless round in his sleep.[20] Nor was it an entirely routine task. Sometimes a fiercer and sometimes a gentler fire was required, depending on whether founding or working was in progress. The *tiseur* of the box is down on his knees anxiously studying the state of his fire.

So we pass round to the front of the box (Fig. 1). The glass now ready to work, here are the glassblowers, the assistant on the right clad in green breeches, bright blue tunic, and green cap, the master on the left seated in his long-armed chair wearing a light blue jacket and a brighter blue cap. The servitor has already blown the "first post," and the paraison is rendered in opaque-white enamel, the only instance of an opaque enamel on the box. The master is trundling his pontil on the arms of his chair,[21] finishing off a drinking glass which is represented in translucent emerald enamel on the gold. Little needs to be said about this aspect of the art, about which so much has been written. Behind them, in an ogee-headed niche, are two biconical drinking glasses, one on either side of a narrow-necked flask, a feature which is less plausibly represented also on the back of the box (Fig. 4).

Very curious is the fact that the gaffer is working with his pucellas held in his right hand while with his left he revolves the pontil, holding the glass on his left side, a remarkably awkward mode of work which one may venture to think the artist never saw in a glasshouse. Perhaps he decided that to make the glass project on the outside of the box was to invite injury to it. It was a sufficient problem to fit a chair and a man and an iron rod on the narrow ledge offered by the flange of the base.

Progressing in our tour, we come on the second narrow side to the "Sarole man," who is either putting in, or taking out, a spherical-bodied flask which is visible in the opening of the lehr (Fig. 8). Since normally the finished glass was taken to the lehr by a boy using a forked stick, we may suppose that this man is taking a glass out at the cooler end of the "tunnel."[22] The hotter end is presumably represented at the other extremity of the furnace, above the "teasehole" man's head. Is it by coincidence that the framework of the opening is given a three-dimensional effect by slanting hatch-lines, suggesting the tunnel-like access

18. *Recueil de Planches* . . . , X, Paris, 1772, pls. 1 and 2 of article "Verrerie en Bois . . . Petite Verrerie à Pivette."

19. J. E. Anderbjörk, ed., *Glasbruksminnen berättade av Svenska Glasarbetare*, Växjö, 1944, p. 41.

20. *Ibid.*, pp. 41–43.

21. For a discussion of the introduction and significance of the glassmaker's chair, see R. J. Charleston, "Some Tools of the Glassmaker in Medieval and Renaissance Times, with Special Reference to the Glassmaker's Chair," *Glass Technology*, 3, no. 3, June 1962, pp. 107–111. For a chair of a somewhat similar model in a late seventeenth-century English glasshouse, see John Burroughs' trade-card, ill. R. J. Charleston, "Pottery, Porcelain and Glass," *The Stuart Period, 1603–1714* (*Connoisseur Period Guides*), London, 1957, p. 92; see also Fig. 11 here.

22. Cf. Fig. 4 of the preceding article, where the "Sarole man" stands nonchalantly on the end of a bench waiting for the next thing to do. He carries no implement. For a boy carrying a glass on a stick, presumably to the lehr, see Fig. 9 here and Fig. 28 of the previous article; for another, using a forked stick, see Diderot, *Encyclopédie, Recueil de Planches* . . . , X, pl. 20 of article "Verrerie en bois. . . ."

to the lehr? This workman is dressed in blue cap and breeches and a green tunic from which much of the enamel has fallen away.

The details of the glassworking on the whole ring true. What of the furnace? Here we come up against an immediate difficulty. If the arched openings at the chamfered corner of the structure are to be taken as the working holes, as seems most likely,[23] then processes requiring two differing degrees of heat are being carried on in the same chamber of the furnace. We must therefore suppose that the artist has taken the liberty of blending two furnaces in one, either combining on one level the two stories of a Venetian-style furnace (which seems rather unlikely, since the furnace is basically rectangular in plan), or depicting a founding furnace if you are looking at the broader sides of the box, and an annealing furnace if you look at the narrower. This latter hypothesis seems to point very strongly in the direction of the "German"-type furnace which is to be seen in its perfected form in Johann Kunckel's furnace (see Figs. 21–22 of the previous article). It is in this sphere that the only basically rectangular structures with chamfered corners are to be found—at Trestenshult, in Sweden; Stenhule, in Denmark; and Junkernfelde, in eastern Westphalia, the two first-named, although lying outside German territory, being almost certainly run by German masters. Mogens Schlüter, in his commentary on the Stenhule furnace, points out that chamfered corners on an otherwise rectangular furnace would give easier access to three pots on the siege in a six-pot furnace.[24]

The interpretation of the upper portion of the box, represented by the cover, is even more fraught with uncertainty (Figs. 2, 4, etc.). On the assumption that it represents the upper part of the furnace, as distinct from the roof of the glasshouse, we are confronted with the difficulty presented by the four arched louvers and the central chimney-like structure. There is no furnace known to the writer which has

FIG. 8. *Side view of the Houghton box, showing* Sarole *man.*

both these features. Several furnaces have chimney-like finials rising above them, but none of them are shown belching out smoke,[25] which issues, where represented at all, through the working holes. The whole object of the furnace was to make the heat circulate round and over the pots on the siege, and not rush to waste up through a chimney in the roof.[26] The arched louvers, however, are seen in the upper vault of the furnace represented as

23. Working-holes are quite often represented in the iconography as being arched openings rather than square or circular holes, the latter requiring special refractory "port-holes" (see n. 16 of previous article). For the arched "bocca," see *ibid.*, Fig. 3 and notably Fig. 21, Kunckel's own furnace.

24. See p. 26 and notes 44, 46 of the previous article. The Trestenshult and Junkernfelde furnaces, however, appear to have housed four pots. By a curious inversion, the French furnace of this type employs the circular furnace for the founding and the rectangular for annealing (Diderot, *Encyclopédie, Recueil de Planches* . . ., X, Supplementary pls., "Verrerie, . . . Four rond ou à la Française"), in line with Agricola's "second furnace" with annexed annealing furnace (Fig. 24 of previous article).

25. See the Preussler furnace shown in Fig. 25 of the previous article, and the Reichenau furnace illustrated here in Fig. 9.

26. See p. 28 of the preceding article, and Fig. 10 here.

the frontispiece to G. L. Bianconi's *Due Lettere di Fisica* (Fig. 11 of the preceding article). The exact structure of this, however, is uncertain and the engraving is open to suspicion as a faithful representation of any existing furnace.[27] It seems just possible that comparable openings occurred in a three-story furnace as access holes to the "tower" of the lehr, and that the artist of the box showed them as features of the furnace-dome although he had already represented the annealing activities of the "Sarole" man at the lower level of the working holes.

On the other hand, if the cover of the box is accepted as the roof of the glasshouse itself, one is faced with the difficulty that it is domed in structure and that the dome and chimney appear to be constructed of bricks or stones. Until the introduction of the English glass cone, it is virtually impossible to trace a glass furnace which was built within a solid vaulted building, whether in brick or stone. The roof is invariably a sloping affair of timber and shingles, even if there is a wall (sometimes solid enough) to support this often very extensive structure.[28] Thus, even in the case of the Reichenau furnace, although it appears to stand in a solidly constructed chamber, in fact the configuration of the glasshouse itself, seen in the upper part of Fig. 9 here, shows a great hipped roof with smoke rising from a central stack. The same pattern may be observed in later glasshouses of Germanic origin, and the frontispiece to the Augsburg Bible of 1730 (Fig. 10) shows the smoke issuing from the working holes of the founding-furnace and the apertures of the lehr, and then rising to emerge finally through a central louvered lantern.[29] This is obviously also the pattern of the Kunckel glasshouse (Fig. 21 of the previous article), although the roof is represented in a curious convention which seems to leave the furnace and the glassblowers out in the open!

FIG. 9. *Engraving by Clemens Beuttler, showing view of Reichenau, with glasshouse (4), and, below, an interior view of it, showing furnace (1), lehr (2), and spreading furnace (3) (?). Mid-seventeenth century.*

In short, the exact form of the furnace represented by the Houghton box cannot be interpreted in all its details. One is forced to accept a measure of fantasy in the formation of a toy such as this—acanthus leaves do not normally grow up the angles of glass furnaces and wanton on the roof. All that can be said is that its main features seem to indicate a "German" furnace of late seventeenth- or perhaps early eighteenth-century date.

One further detail may possibly help to throw light on the origins of the box—the costume of the workmen. The conical caps, in some instances more pointed than in others, seem a constant feature, but apparently not one

27. See n. 24 of the preceding article.

28. For example, in Diderot's *Encyclopédie* all the glasshouses except the "Verrerie Angloise" (those for "petite verrerie," for crown glass, for plate and for bottles) are covered with roofs of this type. Roofs of this kind are also normally seen on glasshouses within the German sphere; see H. Kühnert, "Eine Erbteilung auf der Glashütte Glücksthal bei Lauscha (Thüringen) im Jahre 1809," *Glastechnische Berichte*, 20. Jahrg., H. 7, 1942, Fig. 1; the Nyby glasshouse in Finland, 1801—see Jacob Seela, "The Early Finnish Glass Industry," *JGS*, 16, 1974, Fig. 2.

29. This construction is seen as late as 1777 in London; see Fig. 17 of previous article.

FIG. 10. *Engraving by Johann Georg Pinz illustrating Chap. 33 of the Book of* Deuteronomy, *from the Augsburg Bible of 1730. The Corning Museum of Glass.*

which is distinctive of any particular area of glassmaking. In the illustrations of the great French *Encyclopédie*[30] the men normally wear a cap except when they are doing particularly hot work, such as pot-changing, when they adopt a wide-brimmed felt hat.[31] Their normal costume is a shirt tucked into the top of their breeches, except that again, when working on particularly hot jobs, they wear a long smock which falls straight from the shoulder. In Bohemia, the workmen, as represented on the Preussler *Humpen* of 1680, wear a rounded cap and shirt tucked into rather baggy breeches (Fig. 25 in the previous article); on the 1727 *Humpen* they seem to have gone over (with one exception) to flat-topped caps rather like that worn by Tenniel's *Carpenter* in *Alice in Wonderland* (see Fig. 27 of previous article). In Kunckel's illustration both the workers seem to wear brimmed hats turned up at the front: one has his shirt tucked into his breeches, while the other has a smock-like tunic belted in at the waist in a manner reminiscent of those working on the box. Perhaps the nearest parallels to the glassworkers of the Houghton box are provided by the frontispiece to the Augsburg Bible (Fig. 10), where the belted smock, the pixie cap, and the boots are standard dress, although this is unfortunately less clear in the representation of the glassworkers than it is in that of the carriers. In a late seventeenth-century German engraving illustrating "The Glassmaker" (Fig. 11), however, the cap, the belted smock, and the long boots are all in evidence. All in all, the German parallels seem to be the most compelling, and strengthen the general impression that the glasshouse "represented" by the Houghton box is a German one from the late seventeenth or early eighteenth century.[32]

FIG. 11. *Print of "The Glassmaker" by Kaspar Luyken, 1695.*

30. *Recueil de Planches . . .*, X, *passim.*

31. The felt hat seems to be characteristic of glassmakers in the Venetian tradition—see the frontispiece to the 1669 Amsterdam edition of A. Neri's *L'Arte Vetraria* (Fig. 7 of previous article, and cf. Fig. 5 here). The English workmen apparently wore wide-brimmed straw hats for the hottest types of work "compell'd to work in their shirts like Cyclopes with a straw broad-brim'd Hat on their heads to defend their eyes from excedency of heat and light" (C. Merrett, *The Art of Glass*, London, 1662, p. 248).

32. It should perhaps be added that the glasses themselves which are represented on the box are too generalized in shape to give any clue to a country of origin, although the fact that they are mostly enameled in green might point to the *Waldglas* of German furnaces rather than to the *cristallo* of the Venetian tradition. A further point in favor of a German or Central European origin for the box is the presence of the faceted bottles, wheel-cutting being a specialty of the Germanic countries and not a technique to be expected in the sphere of *façon de Venise* glassmaking until well on into the eighteenth century.

LAMPENGEBLASENES GLAS DES HISTORISMUS
die Hamburger Werkstatt C. H. F. Müller

Helmut Ricke

Die Geschichte des lampengeblasenen Glases ist noch weitgehend ungeschrieben. Sicher hat die Vorstellung, es handle sich bei den Arbeiten in dieser Technik lediglich um Figürchen, Spielzeug, Perlen und Nippes, einer eingehenderen Beschäftigung im Wege gestanden. Erst durch die Werke des Berliner Kupferstechers und Chemikers Karl Koepping, der mit seinen Jugendstil-Ziergläsern seit 1895 weltweite Beachtung fand,[1] scheint das lampengeblasene Kunstglas jenes qualitative und künstlerische Niveau erreicht zu haben, das das Interesse des Kunsthistorikers rechtfertigt. Dementsprechend setzte die Forschung bei der Zeit um 1900 an.[2]

Doch wie bei vielen der scheinbar so eigenständigen Werke des Jugendstils stösst man bei der Frage nach dem Woher der fragilen Schöpfungen Koeppings auf Voraussetzungen im Historismus der zweiten Hälfte des 19. Jahrhunderts. Dem forcierten Bemühen dieser Zeit um die Belebung der Gewerbe und ihrer handwerklichen Grundlagen verdankte die traditionelle Lampenbläserei wesentliche neue Impulse. Die Erweiterung der bisher eng begrenzten Möglichkeiten dieses Kunsthandwerkszweiges zeigte sich vor allem in der Auseinandersetzung mit dem künstlerischen Hohlglas, einem Bereich, der sich bisher aus technischen Gründen der Arbeit vor der Lampe weitgehend entzogen hatte. Man griff Pokal-, Stengelglas- und Vasenformen vergangener Epochen auf und suchte sie in der für die Lampenarbeit charakteristischen Weise aus vorgefertigten massiven Glasstäben oder Röhren nachzubilden, zu variieren und schliesslich zu übertreffen. Auf die technischen Voraussetzungen für diese Ausweitung des traditionellen Repertoires wird im folgenden zurückzukommen sein.

Wie in den übrigen Bereichen des Kunsthandwerks waren es in erster Linie Renaissancevorbilder, die für die Kunstglasbläser des Historismus wirksam wurden, vor allem venezianische Gefässe des 16./17. Jahrhunderts; deren Fertigungstechnik bot sich hierfür an, da sie, analog zur Arbeit vor der Lampe, in reiner Formarbeit und ohne nachträgliche Veredelung gestaltet wurden.

Bekannt ist, dass der zunächst in Lauscha, dann in Wiesbaden tätige Friedrich Zitzmann,

1. 1900 Grand Prix auf der Pariser Weltausstellung. Zum gegenwärtigen Forschungsstand vgl. Helga Hilschenz, *Das Glas des Jugendstils, Katalog der Sammlung Hentrich im Kunstmuseum Düsseldorf*, Düsseldorf 1973, S. 96 ff.

2. Grundlegend wurden die dem lampengeblasenen Glas gewidmeten Kapitel der Dissertation von Clementine Gräfin Schack von Wittenau, *Glas zwischen Kunsthandwerk und Industrie-Design, Studien über Herstellungsarten und Formtypen des deutschen Jugendstilglases*, Köln 1971, S. 38–45, 68–91.

bevor sein Schaffen durch den Anstoss Koeppings in die neue Richtung des Jugendstils gelenkt wurde, Ziergläser im venezianischen Stil gearbeitet hat, die in den 90er Jahren auf der Weltausstellung in Chicago und bei verschiedenen anderen Anlässen Interesse fanden.[3] Die Voraussetzungen dieser frühen Werke Zitzmanns, die Anregungen der venezianischen Glaskunst in freier Form verarbeiten,[4] sind bisher in der Literatur unerörtert geblieben. Zum einen ist hier natürlich die Wiederbelebung der venezianischen Glaskunst in Murano selbst durch die Initiative Antonio Salviatis zu nennen.[5] Doch auch im Bereich des lampengeblasenen Glases gab es lange bevor Zitzmann sich dem Blasen von Ziergläsern zuzuwenden begann, zumindest einen Künstler, der sich dem Glas im venezianischen Stil intensiv widmete—der aus Thüringen stammende Carl Heinrich Florenz Müller in Hamburg. Die bisherige Auffassung, derzufolge Zitzmann die Glasbläserei vor der Lampe für das Gebiet des künstlerischen Hohlglases neuentdeckt habe,[6] muss entsprechend korrigiert werden.

Müller war Zitzmann, wenn nicht im Künstlerischen, so doch im Handwerklich-technischen deutlich überlegen. Dem entspricht seine weitere Entwicklung: die eigentliche Leistung und historische Bedeutung C. H. F. Müllers liegt nicht auf dem Gebiet der Glasgestaltung sondern auf dem der medizinischen Technik; 1895 baute er eine der ersten Röntgenröhren in Deutschland und wurde damit zum Gründer eines noch heute existierenden Weltunternehmens für Durchleuchtungsapparaturen.

Am Anfang des beruflichen Weges dieses erstaunlichen Mannes steht jedoch seine Tätigkeit als Kunstglasbläser, der er von 1865 bis in die zweite Hälfte der 70er Jahre den Hauptteil seiner Arbeitskraft gewidmet zu haben scheint. In der Fachliteratur wurde sein Name bisher nur am Rande erwähnt.[7] Mehr zur Person C. H. F. Müllers tragen einige schwer zugängliche Publikationen bei, die das von ihm begründete Unternehmen anlässlich seines 100-jährigen Bestehens und zu Werbezwecken herausbrachte;[8] sie konzentrieren sich allerdings vornehmlich auf Müllers Verdienste um die Entwicklung der Röntgentechnik. Aus den dortigen Hinweisen und sonstigen Quellen ergibt sich folgendes Bild: Am 29. Januar 1845 wurde Carl Heinrich Florenz Müller in dem nahe Lauscha gelegenen Piesau im Thüringer

---

3. Zu Zitzmann vgl. Schack (s.Anm.2), S.38 ff. und S.168, Anm.4; vgl. auch Hilschenz (s.Anm.1), S.129 ff.

4. Es sind keine Gläser Zitzmanns bekannt, die unmittelbar auf altvenezianische Vorbilder zurückgehen; eine Studie des Verfassers über die historisierenden Gläser Friedrich Zitzmanns ist in Vorbereitung.

5. Vgl. Barbara Mundt, *Historismus, Kunsthandwerk und Industrie im Zeitalter der Weltausstellungen.* Kataloge des Kunstgewerbemuseums Berlin, Bd.VII. Berlin 1973, Kat. Nr. 23, 129–153, 164/65 und Text in Anhang. Der grosse Erfolg der Ziergläser der Società Salviati auf der Pariser Weltausstellung des Jahres 1867 ist symptomatisch für das Interesse der Zeit.

6. Vgl. Schack (s.Anm.2), S.38.

7. Vgl. einen unbetitelten Aufsatz über Koepping in *Kunstgewerbeblatt*, Neue Folge, 8.Jg., 1897, S.163, dort versehentlich unter dem Namen D. F. C. Müller (Verwechslung mit einer bedeutenden Hamburger Kunstschreinerfirma). Auch Edmund Meier-Oberist, *Das neuzeitliche hamburgische Kunstgewerbe in seinen Grundlagen, Ein Beitrag zur Kulturgeschichte des 19.Jahrhunderts*, Hamburg 1925, S.275, erwähnt ihn nur beiläufig. Das gleiche gilt für Schack (s.Anm.2), S.44, Anm.33. Vgl. ferner Axel von Saldern, Originals-Reproductions-Fakes, in: *Annales du 5e Congrès de l'Association Internationale pour l'Histoire du Verre*, Liège, 1972, S.312 f.mit Abb. und Ausst.Kat. *Hohe Kunst zwischen Biedermeier und Jugendstil: Historismus in Hamburg und Norddeutschland*, bearbeitet von Hermann Jedding und Mitarbeitern, Museum für Kunst und Gewerbe Hamburg 1977, Kat. Nr. 327/28 und S.477.

8. Den Hinweis auf die Existenz der Firma Müller verdanke ich Frau Anna-Elisabeth Theuerkauff-Liederwaldt, Berlin, die Möglichkeit der Einsichtnahme in die Firmenschriften auf Vermittlung von Herrn Hermann Jedding, Hamburg, dem freundlichen Entgegenkommen der C. H. F. Müller GmbH Hamburg, vertreten durch die Herren Opeldus und Eilken.

Folgende Schriften ergeben Hinweise: Carl Heinrich Florenz Müller, ein Glasbläser erlangt Weltruf, in: *Röntgenstrahlen, Mitteilungen aus der Röntgentechnik und ihren Nachbargebieten*, hrsg. von der C. H. F. Müller GmbH, Nr. 14, 1965, S.39 ff.; *Ein grosser Name in der Röntgengechnik. C. H. F. Müller.* Hamburg 1971; Heinz Todtmann, *So sehen und wissen wir mehr*, hrsg. von der C. H. F. Müller GmbH, Hamburg 1956, S.11 f.; Aus der Geschichte des Hauses C. H. F. Müller, in: *Wir bei Philips* (Werkszeitschrift), Heft IX/X, 1965, S.6 ff.

Wald geboren, einem Dorf, für das, wie für die meisten Ortschaften dieser Region, das Glasbläserhandwerk noch heute der wichtigste Erwerbszweig ist. Er entstammte einer alteingesessenen Glasbläserfamilie; sein Vater, Johann Christian Gustav Müller, war als freischaffender Lampenbläser tätig, sein Grossvater, Johann Nicol Müller, als Glasmeister der Pisauer Hütte.[9] Wenn auch die Möglichkeit einer zunftmässigen Lehre in den Heimarbeitsbetrieben des Thüringer Waldes nicht gegeben war—man lernte in der väterlichen Werkstatt oder bei Verwandten—so muss der junge Florenz Müller doch bereits in seinem Handwerk voll ausgebildet gewesen sein, als er sich 1863 als 18 jähriger in Hamburg niederliess.

Man wird davon ausgehen können, dass Müller, bevor er sich 1865 selbständig machte, zwei Jahre in der Glasbläserei C. P. Greiner arbeitete, die seit 1863 in Hamburg als Spezialwerkstatt für medizinische und wissenschaftliche Glasgeräte zu belegen ist.[10] Der Name Greiner verweist wiederum auf den Thüringer Wald. Die Greiner gehören wie die Müller zu den weitest verzweigten der eingesessenen Glasmacherfamilien der Lauschaer Region. Vermutlich hat C. P. Greiner den jungen Müller nach Hamburg geholt, nachdem er zunächst allein gearbeitet hatte.[11] Einen Beleg für den Zusammenhang zwischen Greiner und Müller bietet die Lage der Greiner'schen Werkstatt am Neuerwall 74. Im Nachbarhaus, Nr. 72, wohnte die Familie von Dorum; ein Mitglied dieser Familie, Emma Maria Ludowiecke von Dorum, wurde 1866 Müllers Frau.[12]

Wie die Beziehungen zwischen den beiden Thüringer Glasbläsern sich im einzelnen gestalteten und was schliesslich zur Trennung Müllers von Greiner führte, ist heute nicht mehr feststellbar. Sicher ist auf jeden Fall, dass Müller sich bei seinem Schritt in die Selbständigkeit bereits einem leistungsfähigen Konkurrenzbetrieb gegenübersah. Es lag daher für ihn nahe, sich auf ein Gebiet zu verlegen, mit dem sich Greiners Werkstatt nicht befasste, dem der künstlerischen Glasgestaltung. Einen ersten Beleg hierfür liefert Müllers Gewerbe-Anmeldungs-Schein, den er sich am 30. Mai 1868, drei Jahre nach der Gründung seiner Werkstatt, ausstellen liess. "Glaskünstler" ist dort als Berufsbezeichnung angegeben.[13]

Den Einträgen in die Hamburger Adressbücher ist zu entnehmen, wie sich Müllers Betrieb aus kleinen Anfängen entwickelte. 1867–69 ist dort unter der Anschrift Alter Steinweg 43 schlicht "C. H. F. Müller, Glasbläserei," zu lesen. 1870/71 bezeichnet Müller sich als "Fabrikant künstl. Glasbläserei"; ein Umzug innerhalb der gleichen Strasse—die Adresse lautete nun Alter Steinweg 30 u. 32—hatte für den florierenden Betrieb mehr Platz geschaffen. Von 1873 bis 1876 lautet der Eintrag "Fabrik und Lager künstl. Glasbläserei, chirurg., chem., physikal. Instrumente u. Geräthschaften. Speersort 20." Die abermalige Verlegung des Betriebes diente offensichtlich in erster Linie der Erweiterung der Produktion in Richtung der bisher vorwiegend von der Greiner'schen Firma hergestellten

9. Ch. G. Müller, Handarbeiter, von Pisau, *21.10.1817, †12.9.1901, verehelicht mit Johanne Katharina Elisabeth Müller, geb. Rost von Spechtbrunn, *27.8.1824, †21.1.1888. Die Auszüge aus den Pisauer Kirchenbüchern verdanke ich Herrn Pfarrer Wolff, Pisau, und der freundlichen Vermittlung von Frau H. Wallstab, Darmstadt.

10. Seit diesem Jahr ist die Firma als Glashütte für techn. Apparate im Hamburger Adressbuch verzeichnet. Die Aufzählung der Produkte der Werkstatt auf S. 15/16 des Katalogs der Hamburger Gewerk-Ausstellung 1863 macht deutlich, dass es sich dabei ausschliesslich um lampengeblasene Gerätschaften für den medizinischen Bedarf und um Labor- und Forschungsinstrumentarium gehandelt hat. C. P. Greiner war der einzige Betriebs, der derartige Glasgeräte ausstellte.

11. Dies würde erklären, wieso die Firma C. P. Greiner erst ab 1863 im Hamburger Adressbuch aufgeführt ist. Denkbar wäre auch, dass Greiner und Müller gemeinsam nach Hamburg kamen.

12. Mitteilung des Staatsarchivs Hamburg. Das Ehepaar Müller hatte drei Kinder, einen Sohn und zwei Töchter. 1874 wurde Müller in Hamburg eingebürgert. Im gleichen Jahr erfolgte die Eintragung seiner Firma ins Handelsregister.

13. Original im Archiv der Firma C. H. F. Müller.

Geräte.[14] Seit 1877 werden die künstlerischen Arbeiten im Adressbuch nicht mehr erwähnt. Der Eintrag lautet nun—seit 1878 unter der Anschrift Sonninstr. 23—"Glasinstrumenten-Fabrik, Fabr. u. Lager chirurg., chemisch., pharmaceut., physical. u. metereolog. Instrumente u. Geräthschaften v. Glas." Dabei bleibt es bis 1887. Der Aufschwung der Firma C. H. F. Müller fällt sicher nicht zufällig zusammen mit der Auflösung des Greiner'schen Betriebes, der seit 1879 nicht mehr in den Adressbüchern verzeichnet ist.[15]

Konkrete Hinweise auf die von Müller hergestellten Kunstgläser lassen sich erst relativ spät finden. 1876 wurde er auf der grossen *Ausstellung Hamburgischer Industrie-Erzeugnisse* dieses Jahres "für sehr geschickt ausgeführte Glasbläser-Arbeiten, u. a. von venezianischen Flügelgläsern, wie sie in Hamburg bisher noch nicht gefertigt wurden," mit einer Silbermedaille ausgezeichnet (Verzeichnis der Preisträger im Katalog der Ausstellung). Der Katalogeintrag lautet: "C. H. F. Müller, err. 1864 [sic]: Imitation altvenezianischer Flügelgläser mit Kneifzangenarbeit (stellt sonst wissenschaftliche Geräte her)."[16] Müller scheint auf dieser grossen Gewerbeausstellung seine Ziergläser erstmals dem breiteren Publikum vorgestellt zu haben. Auch auf der III. Deutschen Kochkunstausstellung, die 1880 in Hamburg stattfand, war er mit seinen Ziergläsern vertreten.[17]

Welch hohen Rang er selbst diesen Arbeiten beimass, belegt die Tatsache, dass er 1877 eine Kollektion seiner Flügelgläser Kaiser Wilhelm I. als Geburtstagspräsent überreichen liess. Dies brachte ihm ein Anerkennungsschreiben des Königlich Preussischen Hof-Marschall-Amtes ein, in dem ihm—verbunden mit "Allerhöchstihrem besonderen Dank"—werbewirksam bestätigt wurde, "dass seine Majestät durch die . . . in so freundlicher Weise als Geburtstagsgabe eingeschickten und selbstgefertigten imitierten venetianischen Gläser um so mehr erfreut worden ist, als durch dieselben der Beweis von hoher Vollkommenheit vaterländischer Industrie geliefert wird, deren Fortgang . . . wohl zu wünschen ist."[18]

Zu Beginn der 80er Jahre scheint die Herstellung von Ziergläsern für Müllers Werkstatt neben der Fertigung naturwissenschaftlicher Instrumente noch eine bedeutende, wenn auch nicht mehr die führende Rolle gespielt zu haben. Doch schon 1874 hatte er mit der Produktion elektrischer Gasentladungsröhren nach Geissler, Crookes und Hittorf begonnen, darunter vor allem spielerisch gestaltete Zierobjekte, die mit geheimnisvoll-fluoreszierendem Leuchten die Wirkung elektrischer Spannung auf gasbeschickte Vakuumröhren demonstrierten (vgl. Abb. 25).[19] Die Möglichkeiten der Elektrizität begannen Müller seitdem zunehmend zu faszinieren. Ab 1882 produzierte er, nachdem Th. A. Edison im Jahr zuvor seine bahnbrechende Erfindung in Europa bekanntgemacht hatte, die vermutlich ersten und über Jahre hinaus anerkannt besten Glühlampen in Deutschland. Es folgte die Errichtung einer eigenen Blockstation zur Erzeugung des Stroms für die von ihm gelie-

14. Der Adressbucheintrag der Firma Greiner lautet zwischen 1867 und 1878 unverändert "Fabr. u. Lager chem., optisch u. physik. Apparate. Fabrik künstl. Augen."

15. 1879 fehlt der Eintrag völlig. Ab 1880 ist "Greiner, Paul, Fabrikant künstl. Augen" verzeichnet. Entweder ist dies ein Sohn C. P. Greiners oder es handelt sich um C. P. Greiner selbst, der sich in diesem Fall, möglicherweise in Absprache mit Müller, ausschliesslich der Herstellung von Glasaugen zugewandt hätte.

16. Kat. S.22. Dem *Führer durch die Ausstellung hamburgischer Industrie-Erzeugnisse*, Hamburg 1878, S.19 ist zusätzlich zu entnehmen, dass Müller zu je ½ für den deutschen Markt und für die überseeische Ausfuhr arbeitete. Auf die 1876 gezeigten Gläser nimmt Justus Brinckmann in der ersten Folge einer Artikelserie im *Hamburger Correspondenten*, Nr. 219 vom 14.Sept.1879 Bezug: "Die Glasbläserei zeigte, wenn auch nur auf dem beschränkten Gebiet der Nachbildung altvenezianischer Flügelgläser, einen frischen Anlauf."

17. Ausst. Kat. Glas, S.7/8.

18. Vgl. *Wir bei Philips*, (s.Anm.8), S.11.

19. In dem Führer für die Gewerbeausstellung von 1876 (s.Anm.16) werden die von Müller gezeigten *Geissler'schen Röhren (in Tätigkeit)* eigens hervorgehoben.

ferten Anlagen.[20] Die Herstellung von Ziergläsern trat im weiteren Verlauf der 80er Jahre mehr und mehr in den Hintergrund. Auf der grossen Hamburger Gewerbe-Ausstellung des Jahres 1889, auf der Müller eine Goldmedaille errang, hatte er keine Kunstgläser mehr in seinem Angebot.[21]

Mit dem Bau einer der ersten Röntgenröhren—bereits wenige Wochen nach der Publikation der aufsehenerregenden Erfindung des Würzburger Wissenschaftlers—war Müller dann endgültig der Weg gewiesen, der seine Firma zu ihrer heutigen Bedeutung führen sollte. Müller selbst hat sein Unternehmen nur noch bis 1904 leiten können. Am 24.November 1912 starb er an den Spätfolgen der Strahlenschädigungen, die er sich bei seinen röntgentechnischen Experimenten zugezogen hatte.[22]

Die Tatsache, dass Müller sich mit seinen venezianischen Gläsern erst zu einer Zeit an die Öffentlichkeit wandte, als diese Gläser bereits nicht mehr zu seinen wichtigsten Produkten gehörten, ist nur scheinbar ein Widerspruch. Zum einen ist natürlich davon auszugehen, dass seine Firma in den Aufbaujahren zu klein war, um die kostspielige Beteiligung an den grossen Ausstellungen tragen zu können. Hinzu kommt jedoch ein zweiter Punkt, der Müllers Tätigkeit etwas im Zwielicht erscheinen lässt. Als Begründung für den Ankauf einer Gruppe von 7 Gläsern, die 1893 aus Restbeständen der Müller'schen Produktion in das Hamburger Museum für Kunst und Gewerbe gelangten, vermerkt Justus Brinckmann in seinem Erwerbungsbericht: "Eine Anzahl Ziergläser in den Formen der venetianischen Flügelgläser des 17.–18. Jahrhunderts sind Erzeugnisse einer hamburgischen Werkstatt unserer Zeit, aus der zahlreiche wohl gelungene, bisweilen täuschende Nachbildungen von alten Flügelgläsern hervorgegangen sind. Zu deren Anfertigung haben fremde Antiquitätenhändler schon vor Jahrzehnten Anregung und Modelle gegeben; viele dieser neuen hamburgischen Gläser sind von ihnen als alte venetianische Arbeiten abgesetzt und selbst als solche in öffentliche Sammlungen gelangt. Da der sehr geschickte Verfertiger dieser hamburgischen Flügelgläser, Herr. C. H. F. Müller, in neuerer Zeit deren Herstellung aufgegeben hat, schien es rathsam, noch bei Zeiten eine kleine Mustersammlung seiner Erzeugnisse für das Museum zu sichern."[23]

An anderer Stelle ist davon die Rede, Müller habe "im Auftrage holländischer Händler alte deutsche Gläser gefälscht, welche an feinen Farbunterschieden, auffällig scharfer Profilierung des Stengels und am Fehlen der Bruchstelle unter dem Boden erkenntlich sind. Auch wurden sie nicht vor dem Ofen, sondern nur vor der Lampe gearbeitet, Blasen und umgeschlagener Fussrand nachgeahmt."[24] Aus diesen Angaben lässt sich schliessen, dass Müller in den ersten Jahren seiner Tätigkeit weniger für den freien Verkauf als in festem Auftrag arbeitete, wobei die betreffenden Händler sich seiner Kunstfertigkeit für betrügerische Geschäfte bedienten. Wenn Müller diesen Vorgängen auch zunächst arglos gegenübergestanden haben mag, so kann ihm auf die Dauer

---

20. Vgl. *Wir bei Philips* (s.Anm.8), S.8.

21. Die Firma C. H. F. Müller existierte zu diesem Zeitpunkt vorübergehend nur noch als Teil der *Glühlampenfabrik und Elektrizitäts-Werke zu Hamburg A. G. Basse & Klentze*, deren Direktor Müller war. Sein eigener Betrieb scheint 1889/90 aufgelöst worden zu sein. (Adressbucheintrag 1890: *C. H. F. Müller, in Liq.*). Müller brachte seine Glühlampenfabrikation in die neue Aktiengesellschaft ein, die sich ausschliesslich mit der Herstellung und Installation elektrischer Anlagen befasste (40 Mitarbeiter). Vgl. den Katalog *Hamburgische Gewerbe- und Industrieausstellung 1889*, S.103 und Anzeige S.119. Die Fusion war nicht von langer Dauer. Bereits 1892 ist im Adressbuch vermerkt: *C. H. F. Müller, Electrotech. Fabr. und electr. Beleuchtungsanlagen, Bremerstr.14.*

22. Vgl. Ehrenbuch der Röntgenologen und Radiologen aller Nationen. *Sonderbände zur Strahlentherapie*, Band XXIII, Berlin und Wien 1937, S.19 ff.

23. Vgl. Justus Brinckmann, Museum für Kunst und Gewerbe Hamburg, Bericht für das Jahr 1893, *Jahrbuch der Hamburgischen Wissenschaftlichen Anstalten*, Bd.XI, Hamburg 1894, S.19.

24. Zitiert bei Meier-Oberist (s.Anm.7), S.275, nach Notizen Justus Brinckmanns.

der Missbrauch, der mit seiner Arbeit getrieben wurde, nicht entgangen sein. Je bedeutender seine Firma wurde, desto unangenehmer musste ihm die Bindung an die dubiosen Händler werden. Der engagierte Einsatz auf der Gewerbeausstellung von 1876 und die folgenden Aktivitäten können so als eine Art Flucht nach vorn gedeutet werden, als Versuch, sich vom Makel der Kunstfälschung zu befreien und für die venezianischen Gläser neue, seriösere Abnehmer zu gewinnen. Es liegt ganz im Sinne dieses Bemühens, dass Müller sich seit 1877 von Stilimitation und Nachbildung löste und sich zunehmend auf die Entwürfe anerkannter Kunstgewerbezeichner zu stützen begann.

Vor dem Hintergrund dieser Gegebenheiten ist das Werk C. H. F. Müllers nicht nur unter dem Aspekt der Erweiterung unserer Kenntnis historistischer Glaskunst von Interesse. Da der weitaus grösste Teil der hier vorgestellten Gläser seit ihrer Entstehung unverändert als niederländische oder deutsche Arbeit des 17. Jahrhunderts gilt, versteht sich diese Studie auch als Beitrag zur Klärung der vielschichtigen Fragen, die der Problemkreis des Glases *à la façon de Venise* nach wie vor aufwirft.

Da das im folgenden diskutierte Material aus annähernd 120 Gläsern besteht, schien es sinnvoll, die Angaben zu den einzelnen Stücken in einem Kataloganhang zusammenzufassen. Er ist in drei Gruppen gegliedert. Die Gläser der Gruppe I lehnen sich eng an Vorbilder des 17. Jahrhunderts an; es sind die Stücke, die Müller selbst als "imitierte venezianische Flügelgläser" bezeichnet hat. Gruppe II umfasst Gebrauchsgläser nach Künstlerentwürfen. Gruppe III schliesslich besteht aus Pokalen in der Art der Ziergläser der Gruppe I, die jedoch in Material, Technik und Detailform den Gestaltungsprinzipien der Trinkgläser der Gruppe II folgen.

Einen Ausgangspunkt für den Versuch, das Werk C. H. F. Müllers zu erhellen, bieten die sieben Musterstücke, die Justus Brinckmann 1893 für sein Museum erwarb. Von den Gläsern ist keines im Original erhalten, von einem existieren jedoch neuere Photographien (Kat. Nr. I 1g; Abb; 1. vgl. auch Abb. 53, Nr. 21).[25] Der Pokal wird im Inventar als "Grosses Doppel-Acht-Glas mit Deckel" geführt und als "neuzeitige Arbeit nach dem Vorbild venezianischer Flügelgläser" charkterisiert. Bemerkenswert ist die beträchtliche Grösse von nahezu einem halben Meter. Im Aufbau entspricht er in fast allen Details den hüttentechnisch gearbeiteten Façon-de-Venise Gläsern des 17. Jahrhunderts. Die dem Schaft eingefügte Doppelacht ist ein geläufiges Motiv, das sich bei zahlreichen Schlangenpokalen der Zeit findet.[26] Nicht übersehen werden sollte jedoch die Abwandlung des gewählten Typs. Ist die doppelte Acht bei den Vorbildern meist einer annähernd dreieckigen oder herzförmig gerundeten Grundform eingefügt, so bemüht Müller sich um eine Differenzierung des Umrisses, dessen Linienführung nun deutlicher an sich windende Schlangen erinnert. Das Bemühen um feinteiligere Durchbildung zeigt sich auch in der Gestaltung der Deckelbekrönung, die in üblicher Weise das Schaftmotiv wiederholt, ausserdem jedoch noch einem kleinteilig gebildeten Vogel Platz bietet.

Das nicht in allen Partien sauber und exakt gearbeitete Stück zeigt die Müller'sche Werkstatt nicht unbedingt von ihrer besten Seite. Es ist verständlich, dass Justus Brinckmann 1893 bei seinem späten Erwerb von Müller-Gläsern keine Auswahl von Stücken der ersten Qualität mehr vorfand und sich mit den Resten des Lagerbestandes begnügen musste. Dennoch gibt dieses bei Müller selbst erworbene Stück die stilkritischen Mittel an die Hand, ihm eine Reihe von Gläsern anzuschliessen, die aufs

---

25. Eine Farbabbildung des Stückes in *Röntgenstrahlen* (s.Anm.8), S.43.

26. Vgl. z.B. den Pokal, niederländisch, 1672, im Kunstgewerbemuseum Prag. Inv. Nr. c.9634. Abb.: Kat. Uměleckoprůmyslové museu Praze, *Sbirky Uzitého Umeni*, Prag 1965, Nr. 37.

ganze gesehen die Qualität der Gläser C. H. F. Müllers in günstigerem Licht erscheinen lassen.

Der Pokaltyp mit der Doppelacht im Schaft scheint zu Müllers gängigsten Modellen gezählt zu haben. Zehn Gläser dieser Art sind zur Zeit nachzuweisen, fünf davon haben ihren originalen Deckel bewahrt. Besonders gelungen sind die Exemplare im Viktoria & Albert Museum, London (I 1a; Abb. 2) in der Sammlung Jerome Strauss (I 1c) und im Gemeentemuseum Arnhem (I 1f).

Von Bedeutung für die Zuweisung weiterer Typen von lampengeblasenen Schlangengläsern an die Hamburger Müller-Werkstatt ist der Doppelacht-Pokal I 1i im Frankfurter Museum für Kunsthandwerk. Er ist Teil der Schenkung Kessler-Kollings und gehört zu einer Gruppe von ehemals 10, heute 8 Flügelgläsern, deren einheitliches Erscheinungsbild ihre Herkunft aus der gleichen Werkstatt wahrscheinlich macht.[27] Zu dieser Gruppe gehören unter anderen zwei Gläser mit einer einfachen Acht im Schaft (I 2b, c; Abb. 3). Ein solches "einfaches Acht-Glas mit Deckel" (I 2a), erwarb auch Brinckmann von Müller; es ist heute nur noch durch den Eintrag im Hamburger Inventar zu belegen. Die Frankfurter Stücke repräsentieren in ihrer exakt symmetrischen, sicheren Durchbildung den beachtlichen Qualitätsstandard der gängigen Produktion der Werkstatt C. H. F. Müller. Von den einfachen Acht-Gläsern, deren Schaftverzierung auf das wohl häufigste Ziermotiv der Schlangenpokale des 17. Jahrhunderts zurückgreift,[28] können ausser den beiden Exemplaren in Frankfurt noch ein Pokal im Musée du Verre in Lüttich (I 2d, ähnlich Abb. 53, Nr. 1) und ein weiterer in der Stiftung Heylshof in Worms (I 2e) mit Müller in Verbindung gebracht werden.

Ebenfalls zum Bestand der Schenkung Kessler-Kollings gehören zwei Pokale mit tropfenförmigem Schafteinsatz (I 3c, d; Abb. 4), deren Verwandtschaft insbesondere mit I 2c (vgl. Abb. 3 rechts) unverkennbar ist; es fehlt lediglich die dort eingefügte Acht. Neu ist dagegen die Gestaltung der Kämme, die die vogelkopfartigen Endungen der Schaftverzierung bekrönen. Sie sind aus übereinandergeschmolzenen blauen Glasstäben gebildet. In gleicher Weise ist ein Glas gestaltet, das 1976 in Köln versteigert wurde (I 3e) und ein weiteres im Dresdner Kunstgewerbemuseum (I 3f) dessen Kämme allerdings verloren sind.

Das gleiche Modell wurde auch mit den geläufigeren runden Kämmen mit Waffelmusterung gefertigt, wie zwei Exemplare im Rijksmuseum Amsterdam (I 3a; Abb. 5) und im Fitzwilliam Museum in Cambridge (I 3b) belegen.[29] Eine Variante mit flötenförmiger Kuppa hat sich möglicherweise in einer Skizze erhalten, die Oskar Rauter, der Direktor der für die historistische Glaskunst so bedeutenden Glashütte in Köln-Ehrenfeld, 1881 auf einer seiner Reisen anfertigte (Abb. 6).[30] Während die Gestaltung der Kämme dieses Stückes nicht mit Sicherheit mit der der Pokale I 3c, d zu identifizieren ist, verweisen die wie Schuppen übereinandergelegten seitlichen Ansätze deutlich auf die technischen Eigenheiten der Arbeit vor der Lampe (vgl. auch Abb. 9).

Gleichfalls aus der Sammlung Kessler-Kollings stammen zwei Pokale (I 4a, b; Abb. 7)

---

27. Vgl. Kat. *Europäisches und aussereuropäisches Glas*, bearb. von A. Ohm, Museum für Kunsthandwerk, Frankfurt/M. 1973, Nr. 186–191, S.88/89.

28. Beispiele in nahezu allen grösseren Glassammlungen. Vgl. z.B. *Kataloge des Kunstmuseums Düsseldorf, Glas 1*, bearbeitet von E. Heinemeyer, Düsseldorf 1966, Nr. 269, 272, 273, S.91 f.

29. Auch die einfache Schaftform des Typs I 3 lässt sich bereits im 17. Jahrhundert nachweisen; vgl. z.B. C. Isings, *Schönes altes Glas*, Hannover 1966, Abb. 1966, Abb. 85 rechts.

30. Die Skizzenbücher Oskar Rauters befinden sich im Besitz des Kunstmuseums Düsseldorf. Rauter sammelte in diesen Büchern vorbildliche Glasformen aus der Vergangenheit und aus neuerer Zeit als Grundstock für das Programm der Ehrenfelder Hütte, deren Produktion seit den 80er Jahren des 19. Jahrhunderts an Qualität und handwerklicher Perfektion von keiner anderen deutschen Hütte übertroffen wurde.

Die in Abb. 6 gezeigte Flöte wurde in dieser oder ähnlicher Form allerdings nie von der Ehrenfelder Hütte hergestellt.

deren stark gewundene Schaftverzierung sich ein weiteres Mal von Motiven herleiten lässt, die im 17. Jahrhundert häufig bei Flügelgläsern begegnen.[31] Das gleiche gilt für den auch zu der Frankfurter Gruppe gehörenden Pokal I 5b.[32] Von diesem Typ hat sich in Lund ein weiteres Exemplar mit Deckel erhalten (I 5a; Abb. 8), das in seiner etwas ungelenken Durchbildung stark an das Hamburger Doppel-Acht-Glas (Abb. 1) erinnert.

Formal eigenständiger und technisch vollendeter ist ein Pokal im Amsterdamer Rijksmuseum (I 6a; Abb. 9), dessen dekorativ verschlungene Schaft- und Deckelverzierung das Motiv des geknäulten Schaftes, das von frühbarocken Gläsern *à la façon de Venise* her geläufig ist,[33] dekorativ in die Fläche projiziert. Ausser dem vollständigen Pokal haben sich noch zwei Deckel von Gläsern dieser Art erhalten (I 6b, c).

Häufiger scheint der Typ I 7 gewesen zu sein, von dem sich gegenwärtig fünf Exemplare nachweisen lassen. Unversehrt geblieben ist lediglich das des British Museum in London (I 7a; Abb. 10). Müller hat diesen Typ nicht nur als Deckelpokal sondern auch als deckellosen Spitzkelch produziert; ein Exemplar verwahrt das Fitzwilliam Museum in Cambridge (I 7e). Eine Variante dieses Glases hat sich in einem von den Acht-Gläsern abgeleiteten Kelchglas im New Yorker Metropolitan Museum erhalten (I 8; Abb. 11).[34]

Entwickelte sich bei den bisher gezeigten Gläsern das Schaftmotiv in der Fläche, so tritt bei den Pokalen des Typs I 9 an dessen Stelle eine räumlich gestaltete Figur aus zwei spiralförmig umeinandergewundenen Doppelstäben. Obwohl viele der Pokale mit flächig aufgebauten Schäften an die handwerklichen Fähigkeiten des ausführenden Glasbläsers weit höhere Anforderungen stellten, ist der optische Eindruck der Spiral-Schlangengläser der von äusserst kompliziert und aufwendig gestalteten Arbeiten, so dass man geneigt ist, in diesen und den davon abgeleiteten Gläsern mit Drachenköpfen (I 10) Spitzenstücke der Produktion der Werkstatt C. H. F. Müllers zu sehen. Zur Zeit lassen sich insgesamt fünf Gläser des Typs I 9 nachweisen. Neben einem nur in graphischer Wiedergabe erhaltenen Glas des Wiener Kunstgewerbemuseums (I 9b; Abb. 12) sind ein diesem sehr ähnlicher, völlig unversehrter Deckelpokal in Hartford, Conn. (I 9a) und ein leicht beschädigtes Exemplar im Amsterdamer Rijksmuseum zu nennen (I 9c; Abb. 13). Die beiden verbleibenden Gläser befinden sich im Musée du Verre in Lüttich und im Nürnberger Gewerbemuseum (I 9d, e).

Vorbilder des 17. Jahrhunderts für die räumliche Schaftgestaltung sind selten. Am nächsten kommt der Müller'schen Lösung ein Glas im Kölner Kunstgewerbemuseum.[35] Im Prinzip verwandt, wenn auch nicht im Detail, ist eine Gruppe von Flügelgläsern im Hessischen Landesmuseum in Kassel, die möglicherweise aus der dortigen Produktion stammt.[36]

Häufiger als die Spiral-Schlangenpokale des Typs I 9 sind Gläser überliefert, die anstelle der Kämme mit einem hohlgeblasenen Drachenkopf und einem stark gewundenen Schlan-

---

31. Vgl. z.B. *Kataloge des Kunstgewerbemuseums Köln, Band 1, Glas*, bearbeitet von B. Klesse und G. Reineking-von Bock, Köln 1973, Nr. 308 ff., S. 155 ff.

32. Vgl. z.B. H. E. van Gelder, *Glas en Ceramiek*, Utrecht 1955, Tf. IX, Nr. 6, 9.

33. Vgl. A.-E. Theuerkauff-Liederwald, *Niederländische Glasformen des 17. Jahrhunderts*, Diss. Freiburg 1964, S. 93 ff.

34. Das Zentralmotiv mit begleitenden Schlaufen zur Rechten und Linken ist bei Gläsern des 17. Jahrhunderts ebenfalls nachzuweisen; vgl. z.B. Kat. Köln (s. Anm. 31), Nr. 316, S. 158.

35. Vgl. Kat. Köln (s. Anm. 31), Nr. 314, S. 157.

36. Ob die Gläser bereits in der ersten Kasseler Weissglashütte um 1583/84 gefertigt wurden, ist noch ungeklärt. Es handelt sich um die Stücke mit der Inv. Nr. B IX a 1/1, (Foto Nr. 101, 110, 122, 736/37, 741, 749). Ihre räumlich-plastisch gebildeten Schäfte sind z.T. durch einen komplizierteren Aufbau als das aus einer anderen Werkstatt stammende Kölner Glas (s. Anm. 35) gekennzeichnet. Ein weiteres Glas dieser Gruppe befindet sich in der Eremitäge in Leningrad (Inv. Nr. P 510), ein anderes in der Sammlung Leopold Blumka; vgl. Ausst. Kat. *Three Great Centuries of Venetian Glass*, Corning 1958, Nr. 83, S. 83.

genschwanz versehen sind.[37] Im Pariser Musée Jacquemart-André hat sich ein leicht beschädigtes Exemplar dieses Typs erhalten (I 10a; Abb. 14), das den engen Zusammenhang mit den Gläsern in Wien, Hartford und Amsterdam belegt. Ein unversehrtes Stück wurde 1930, in dem Glauben, es handele sich um eine gegen 1600 entstandene Arbeit, durch das Hamburger Museum für Kunst und Gewerbe angekauft (I 10b; Abb. 15); die Kenntnis des Werkes C. H. F. Müllers war zu diesem Zeitpunkt in Hamburg bereits in Vergessenheit geraten.[38] Diesem Glas sind ein Pokal im Bonner Landesmuseum (I 10c) und ein von Oskar Rauter skizziertes Stück (I 10d) besonders ähnlich. Eng anzuschliessen ist ein kleineres Exemplar in Lüttich (I 10e). Ebenfalls zu dieser Gruppe gehören zwei Flöten, eine stark beschädigte im Gewerbemuseum Nürnberg (I 10i; Abb. 18) und eine weitere im Prager Kunstgewerbemuseum (I 10k). Sie belegen die rationelle Übertragung erfolgreicher Motive auf andere Gefässtypen durch die Müller'sche Werkstatt.

Drei weitere Flöten im Gewerbemuseum (I 11a; Abb. 19) und im Germanischen Nationalmuseum in Nürnberg (I 11c) sowie im Kasseler Landesmuseum (I 11b) können wohl ebenfalls für Müller in Anspruch genommen werden. Sie weisen als einzige Gläser der Müller-Gruppe asymmetrische Schaftmotive auf. Auch sie gehen auf Typen des 16./17. Jahrhunderts zurück.[39]

Zierpokale mit farbigen Blüten als Schaftmotiv sind aus den südlichen Niederlanden und aus Venedig bekannt. An diese älteren Stücke, insbesondere die im 17. Jahrhundert vermutlich in Lüttich entstandenen Gläser,[40] lehnen sich zwei Pokale im Nürnberger Gewerbemuseum an (I 12a, b; Abb. 20). Zu nennen sind ferner zwei Gläser im Wiener Museum für angewandte Kunst (I 13a; Abb. 21) und im Gemeentemuseum Den Haag (I 13b).[41] Das Wiener Stück wurde gemeinsam mit einem lampengeblasenen Hohlbalusterpokal erworben (I 14a), der auf einen verbreiteten Typus südniederländischer Gläser des 17. Jahrhunderts zurückgreift.[42] Da der Pokal I 13a aufgrund formaler und technischer Übereinstimmungen sehr wahrscheinlich als Arbeit Müllers anzusprechen ist, wird man davon ausgehen können, dass das zweite Wiener Glas ebenfalls aus seiner Werkstatt stammt. Zwei weitere Balusterpokale, der eine im Leipziger Grassi-Museum (I 14b; Abb. 22), der zweite, etwas einfachere, in der Moritzburg in Halle (I 14c), sind dem Wiener Glas anzuschliessen.

Ebenfalls für C. H. F. Müller können versuchsweise einige Gläser in Anspruch genommen werden, die als Schaftverzierung hohlgeblasene Tiere aufweisen. Ein Pokal aus der Gruppe der Schenkung Kessler-Kollings in Frankfurt zeigt ein krokodilartiges Fabelwesen (I 15a; Abb. 23), ein Glas der Sammlung Peill & Putzler, Düren, einen Ziegenbock (I 15b; Abb. 24). Dass es nicht abwegig ist, auch derartige Arbeiten als Produkte der Müller-Werk-

---

37. Die Gläser dieser Gruppe hat—ausgehend von einer Entstehung im späten 16. Jahrhundert—A. von Saldern zusammengestellt; vgl. A.v.S., Spiral-Schlangenpokale à la façon de Venise, *Jahrbuch der Hamburger Kunstsammlungen*, Bd. 17, 1972, S. 67–80. Vgl. auch Theuerkauff-Liederwald (s. Anm. 33), S. 90–92 und A. von Saldern, *Reallexikon zur Deutschen Kunstgeschichte*, Stichwort *Façon de Venise*, Bd. VI, Sp. 1024 ff.

38. Abb. 14 zeigt eine Reproduktion aus: Max Sauerland, *Bericht über die Neuerwerbungen des Jahres 1929/30*, Justus Brinckmann Gesellschaft, Hamburg 1930. Für den heutigen Zustand vgl. v. Saldern (s. Anm. 37), S. 72, Abb. 6 (ohne die ebenfalls erhaltene Deckelkalotte).

39. Vgl. z.B. van Gelder (s. Anm. 32), Tf. VIII, Nr. 3.

40. Vgl. z.B. R. L. Chambon, *Histoire de la Verrerie en Belgique du 11e siècle à nos jours*, Brüssel 1955, Taf. XXIV, Abb. 76/77.

41. Der Typ variiert Façon-de-Venise-Gläser des 17. Jahrhunderts; vgl. z.B. Kat. Köln (s. Anm. 31). Nr. 315, S. 158.

42. Vgl. z.B. W. B. Honey, *Victoria & Albert Museum, Glass, A Handbook* . . . London 1946, Tf. 34 D. — Das Glas der Wiener Sammlung wurde 1874 als vorbildliches altvenezianisches Stück in Band I des Vorlagenwerks *Das Kunsthandwerk* aufgenommen, ein sprechender Beleg für die Verwirrung, die Müllers Gläser bereits in ihrer Entstehungszeit verursachten. Vgl. auch den Parallelfall I 9b, s. unten.

statt anzusprechen, mag die Reproduktion einiger Geissler'scher Röhren aus einem Spezialkatalog der Firma C. H. F. Müller aus dem Jahr 1880 illustrieren (Abb. 25).[43]

In diesen Zusammenhang gehören ferner zwei Gläser mit Amorfigürchen im Schaft—ein Pokal im Musée du Verre in Lüttich (I 16a; identisch mit Abb. 54, Nr. 14) und eine Flöte im Nationalmuseum Stockholm (I 16b; Abb. 26). Sie weisen auch im Motivischen Übereinstimmungen mit den Gasentladungsröhren aus Müllers Werkstatt auf.—Aus dem Rahmen des Gewohnten fällt ein in Corning verwahrter Spitzkelch mit geflügelten kettenhaltenden Schlangen zu Seiten des Schafts (I 17; Abb. 27),[44] dessen Verwandtschaft in der Materialbehandlung und in der Einzelform (vgl. Abb. 11) gleichfalls eine Zuweisung an Müller gerechtfertigt erscheinen lässt (vgl. hierzu Abb. 54, Nr. 24).

Der Amorpokal in Lüttich weist eine möglicherweise nachträglich angebrachte Abschlagnarbe auf, die den Eindruck erwecken soll, das Glas sei am Ofen geblasen. Das gleiche Detail fällt bei einem grossen Deckelpokal des Drachentyps (I 10f; Abb. 16) im Victoria & Albert Museum auf; allerdings ist dem Abriss deutlich anzusehen, dass hier nie ein Hefteisen fest angesessen hat.[45] Das Glas weicht durch die ungewöhnliche Farbzusammenstellung der Fadeneinlagen und Ansätze von den übrigen Stücken des Typs ab, weist sonst jedoch alle technischen Eigenheiten der Müller-Gruppe auf. Es wird sich entweder um eine "Sonderanfertigung" Müllers oder um die Arbeit eines geschickten Nachahmers handeln. Die Möglichkeit, dass sich hier ein Glas des 17. Jahrhunderts—vielleicht sogar das Vorbild für die Müller'sche Gruppe—erhalten haben könnte, scheidet aus, da verschiedene technische Eigenheiten den Londoner Pokal eindeutig als lampengeblasen ausweisen und Gläser dieser Wandungsstärke und Grösse vor der Mitte des 19. Jahrhunderts technisch nicht zu bewältigen waren; darauf wird zurückzukommen sein.

Ein zweiter Fall der Aufnahme Müller'scher Formen und Techniken lässt sich an einem Glas im Kunstgewerbemuseum Dresden aufzeigen. Der dortige Drachenpokal (I 10g; Abb. 17) fällt vor allem durch die mangelnde technische Bewältigung des Materials auf. Formal ist der Typus zwar getroffen, die Verwendung unverträglicher Gläser und mangelnde Erfahrung beim Kühlen bewirkten jedoch einerseits das Zerfliessen der Farbfäden des Schafts, andererseits eine übermässige Spannung innerhalb der gewundenen Glasstränge, die zu zahllosen Brüchen und Sprüngen und zum Abplatzen der Ansätze führte. Laut Inventareintrag wurde dies Glas 1876 zusammen mit einigen anderen (z.B. I 5c), die das Museum grösstenteils 1890/91 weiterverkaufte,[46] über den Amsterdamer Händler Goudstikker er-

---

43. Vgl. Preis-Verzeichnis über elektrische Röhren, Radiometer und Crook'sche Apparate von C. H. F. Müller in Hamburg. Heft 1. Hamburg 1880, S. 5. Original im Besitz der Firma C. H. F. Müller. Einige Röhren und sonstige Objekte in diesem Katalog sind mit grossformatigen Tieren ausgestaltet (Storch, Hirsch, Pferd, Pudel, Kamel).

44. Vgl. vor allem ein verschollenes Glas aus der Dessauer Hütte (?), abgebildet bei A. Demmin, *Das Glas. Dessen Geschichte und Werkweise*, Leipzig 1883, S. 59.

45. Erkennbar ist eine Aufrauhung der Oberfläche, die vermutlich durch Andrücken eines entsprechend präparierten Stempels aus gebranntem Ton oder dergleichen an das erweichte Glas entstand. Die Markierung Könnte auch nachträglich von fremder Hand angebracht worden sein.

46. Es werden noch folgende Gläser aufgeführt: Inv. Nr. 3348/48, "Flügelgläser mit Deckel. u. blauen, gekniffenen Verzierungen, am herzförmigen Mittelstück des Ständers u. d. Wiederholung dessl. auf d. Knopfe." Inv. Nr. 3350 "Flügelglas m. einer gewundenen von gelben, roten und blauen Fäden durchzogenen Schlange im Ständer u. auf d. Deckel." Inv. Nr. 3351 "Flügelglas, die gewundenen Schlangen ohne Kopf u. Schwanz wie beim vorigen, auf der Spitze des Deckels ein blaugrüner Vogel." Inv. Nr. 3352 "Flügelglas ohne Deckel, im Ständer eine lyraförmige, von bunten Fäden durchzogene, aussen mit blauen Ansätzen versehene Verzierung."

Die Ähnlichkeit mit den von C. H. F. Müller geblasenen Typen ist offenkundig. Das lyra-förmige Motiv der Inv. Nr. 3352 dürfte mit Typ I 3 oder Abb. 54, Nr. 12 identisch sein. Das beschädigt erhaltene Glas I 3f gehört nicht zu der von Goudstikker erworbenen Gruppe, dürfte jedoch wegen seiner Abrissnachahmung als eine weitere Fälschung in der Art C. H. F. Müllers anzusprechen sein.

worben. Es ist nicht auszuschliessen, dass Goudstikker einer der von Brinckmann erwähnten niederländischen Händler war, in deren Auftrag Müller zunächst arbeitete. Rein hypothetisch wäre denkbar, dass Goudstikker, nachdem Müller seine Mitarbeit aufgekündigt hatte, die bisher von diesem gelieferten Modelle durch einen anderen, weniger geübten Glasbläser arbeiten liess. Dies würde die technische Unzulänglichkeit des Dresdner Glases erklären. Im übrigen handelt es sich nicht um ein Einzelstück; ein weiteres, sehr ähnliches Glas hat sich im Bestand des Museums für Thüringische Volkskunde in Erfurt erhalten (I 10h).[47]

Im Zusammenhang mit den Dresdener Gläsern stellt sich die Frage, ob nicht möglicherweise noch weitere Lampenbläser die gleichen oder ähnliche Modelle wie die von Müller gefertigten geblasen haben. Die Frage ist umso berechtigter, als beim Vergleich der bisher vorgestellten Gläser spürbare Qualitätsschwankungen und Abweichungen in der Ausführung zu verzeichnen sind. So fällt auf, dass ein Teil der Gläser am Fussansatz—meist in Verbindung mit einer Aufwölbung der Bodenplatte—eine Verdickung in der Art von hüttentechnisch gearbeiteten Stücken aufweist, während der Schaft bei den übrigen Gläsern organisch aus dem Fuss aufsteigt. Die Schaftverdickung als Sicherung der Verbindung des angesetzten Fusses und als Verstärkung für den Ansatzpunkt des Hefteisens ist nur bei am Ofen geblasenen Gläsern sinnvoll. Man könnte also zu dem Schluss kommen, dass bei diesen Stücken das Bestreben, die Arbeiten des 17. Jahrhunderts zu kopieren, deutlicher im Vordergrund gestanden habe, als bei den Gläsern ohne die Verdickung, die ja, da hier der Fussteller unmittelbar aus der Schaftröhre oder-stange aufgetrieben wurde, den Gegebenheiten der Lampentechnik deutlicher Rechnung tragen. Es zeigt sich jedoch, dass beide Lösungen bei nahezu identischen Gläsern der gleichen Provenienz wechselweise angewandt wurden, so bei einem Paar aus der Sammlung Kessler-Kollings (Abb. 4) und bei Gläsern aus der Sammlung Völkner (Abb. 20). Die Fertigung beider Varianten in der gleichen Werkstatt dürfte damit hinreichend belegt sein (vgl. auch Abb. 53, 54). Die Detailabweichungen lassen sich generell damit erklären, dass die meisten der Gläser Müllers mit Sicherheit in Werkstattarbeit entstanden. Man wird davon ausgehen können, dass zumindest seit Beginn der 70er Jahre neben Müller noch zwei oder drei weitere Glasbläser an der Produktion der Ziergläser beteiligt waren.[48] Auch muss man sich klar machen, dass die Arbeit vor der Lampe anders abläuft als die am Ofen. Vor der Lampe werden die einzelnen Teile eines Pokals—Kuppa, Schafteinsatz, Fuss—gesondert vorgefertigt und dann miteinander verschmolzen. Auf diese Weise können bei rationeller Arbeitsweise mehrere Glasbläser am gleichen Stück beteiligt sein. Der Versuch, auf stilkritischem Wege Gläser des Meisters von denen der Gesellen zu scheiden, ist unter diesen Voraussetzungen von vorn herein zum Scheitern verurteilt.

Wie diese Gegebenheiten deutlich machen,

---

47. Ausserdem befindet sich im Kunstgewerbemuseum Dresden das Schaftfragment eines weiteren Spiral-Schlangenglases der beschriebenen Art. Dem Dresdener Inventar ist nicht zu entnehmen, ob die acht 1876 von Goudstikker erworbenen Gläser als altvenezianische Stücke oder als neuere Nachbildungen angekauft wurden. Der Preis von 750, — Mark für die Gruppe ist vergleichsweise hoch; Müller erhielt 1893 von Justus Brinckmann für seine sieben Gläser nur 166, — Mark. Die Tatsache, dass man sechs der acht Dresdener Gläser nach wenigen Jahren weiterverkaufte, spricht dafür, dass man zumindest diese zu dem betreffenden Zeitpunkt nicht mehr für Originale hielt.

48. Von Müllers Mitarbeitern ist nur einer namentlich bekannt. 1896, also lange Zeit nach der Einstellung der Produktion von Ziergläsern, wird ein Glasbläser Becker genannt, der anlässlich des Besuches junger Ärzte in Müllers Werkstatt auf dessen Geheiss "drei schöne Pokale, so gross, dass jeder eine ganze Flasche . . . Weines fasste" vor den Augen der Besucher "aus kräftigem Glasrohr über dem Gebläse entstehen" liess; vgl. Prof. Dr. H. Gocht, Die Gründung des chirurgischen Röntgeninstituts am Allgemeinen Krankenhaus Hamburg Eppendorf, *Beiträge zur klinischen Chirurgie*, Bd.XCII, 1914.

können die Detailabweichungen der einzelnen Gläser durch das unterschiedliche Geschick und die handwerklichen Eigenheiten der beteiligten Werkstattkräfte erklärt werden. Dennoch ist nicht grundsätzlich auszuschliessen, dass das eine oder andere der hier vorgestellten Gläser sich eines Tages als Arbeit einer weiteren, heute noch unbekannten Werkstatt erweist, in der die von Müller produzierten Typen geschickter kopiert wurden, als dies der Verfertiger des Dresdner Glases (Abb. 17) vermochte. Unsere Kenntnisse der Lampenbläserei des 19.Jahrhunderts sind hier für verbindliche Aussagen noch zu lückenhaft. Als weitgehend eigenhändige Arbeiten Müllers wird man nur die Gläser ansprechen können, die in den ersten Jahren seiner Tätigkeit, zwischen 1865 und ca. 1870 entstanden, da er seine Werkstatt in diesen Jahren noch allein oder mit höchstens einem Gehilfen geführt haben dürfte. Bei zwei Gläsern, die sehr früh in öffentliche Sammlungen gelangten, ist die Entstehung in den 60er Jahren zu belegen, dem Deckelpokal des British Museum (Abb. 10)[49] und einem bereits 1867 erworbenem Glas in Wien (Abb. 21). Bei beiden Stücken fallen die steil aufsteigenden, relativ schmalen, in prägnante Spitzen auslaufenden Kämme und die etwas starre, wenig geschwungene Gestaltung der schnabelartigen Fortsätze unterhalb dieser Kämme auf. Allgemein wirken die beiden Gläser vergleichsweise spannungslos und bei aller Feinheit in der Durcharbeitung fast etwas schwerfällig. Bei den späteren Arbeiten wird die Gestaltung der Kämme und Fortsätze differenzierter und die Form allgemein gestraffter.

Diese Beobachtungen ermöglichen die Festlegung einer ungefähren Chronologie der Arbeiten aus Müllers Werkstatt; die Datierungen sind im einzelnen dem Katalog zu entnehmen. Auffallend ist, dass bei den frühen Stücken noch häufig die Neigung zum Experiment festzustellen ist. Es werden—zum Teil am gleichen Stück—Zangen mit unterschiedlichen Musterungen verwendet, etwa Waben, Rauten (I 1a, Abb. 2; I 2b, Abb. 3; I 1b; I 9a) oder einfache Parallelriefen (I 1c, d; I 6c; I 11c; I 13a, b).[50] Auch die Gestaltung der Ansätze wechselt; sie variiert bis hin zur Dreiecksform (I 8; I 13a, b). Technisch ist noch keine völlige Sicherheit erreicht; gelegentlich zerfliesst die Zangenmusterung der Ansätze durch zu starkes Nachwärmen des Glases vor dem Abkühlen.[51] Bei den frühen Gläsern ist für die Gestaltung des Fussansatzes ausnahmslos die der Lampentechnik adäquatere Lösung ohne Schaftverdickung gewählt. Von allen Typen mit Kämmen und Schnabelfortsätzen lassen sich Exemplare nachweisen, deren Entstehung aufgrund der genannten Kriterien bereits in der zweiten Hälfte der 60er Jahre anzusetzen ist. In den 70er Jahren wird dies Programm unter stärkerer Beteiligung von Werkstattkräften fortgeführt. In dieser Zeit entwickeln sich die Kämme zu kräftiger gerunderter Form und die Schnabelfortsätze zu lebhafter geschwungenen, mit winklig gekniffenen Endungen versehenen Gebilden. Es sind zahlreiche Zwischenstufen zu beobachten; gelegentlich finden sich auch frühere und spätere Bildungen am gleichen Stück (z.B. I 1f, I 7c). Die Gestaltung der Kuppa scheint zwischen 1865

---

49. Das Glas gehört zur Sammlung Slade, die 1868 als Schenkung ins British Museum gelangte. Obwohl der Pokal durch seine Grösse und aufwendige Gestaltung zu den bedeutendsten Stücken unter den venezianischen Gläsern der Sammlung gehörte, wurde er nicht, wie die übrigen Gläser, in A. Nesbitts 1871 erschienenem Katalog abgebildet. Seine Beschreibung bildet den Abschluss der Gruppe *Venetian Glass Clear and Colourless*. Hieraus kann entweder geschlossen werden, dass Nesbitt das Glas bereits als neuzeitliche Arbeit identifiziert hatte, oder dass es erst kurz vor der Drucklegung des Katalogs in die Sammlung gelangte. 1931 wurde das Glas als flandrisches Schlangenglas des 17.Jahrhunderts publiziert; vgl. W. King, Flämisches Glas und seine Beziehungen zum englischen Glas, *Pantheon*, Bd.VII, S.158 ff.

50. Bei den späteren Arbeiten wurden ausschliesslich Zangen mit aus Quadraten gebildetem Waffelmuster in zwei verschiedenen Grössen verwendet.

51. Das sorgfältige Durchwärmen des fertigen Glases ist unerlässlich, um die bei der Lampenarbeit besonders ausgeprägte Gefahr der Spannungsbildung zu mildern.

und 1875 von niedrigen, stärker ausgestellten zu höheren, geradwandigeren Kelchformen zu tendieren.

Die Gläser C. H. F. Müllers, die im vorigen Jahrhundert in öffentliche Sammlungen gelangten, wurden von den Zeitgenossen fast ausnahmslos als altvenetianische Werke angesehen. Die einzige nachweisbare Ausnahme ist neben den Hamburger Stücken der Pokal I 7c, der 1886 mit dem Vermerk "wohl hamburgische Imitation eines alten Flügelglases" im Germanischen Nationalmuseum Nürnberg inventarisiert wurde. Besondere Auswirkungen hatte die Frühdatierung des Deckelpokals I 9b. Das heute nicht mehr nachweisbare Glas gehörte zum Bestand des Wiener Kunstgewerbemuseums. In dem Glauben, es handele sich um ein venezianisches Werk des 16. Jahrhunderts, wurde es 1875 als vorbildliche Arbeit dieser Zeit in Band II des Prachtwerkes *Das Kunsthandwerk* aufgenommen (Abb. 12).[52] In dem wenige Jahre später erschienenen Glaskatalog des Museums korrigierte Bruno Bucher, einer der Herausgeber der Reihe *Das Kunsthandwerk*, seine Zuweisung und ordnete das Glas, gemeinsam mit zwei weiteren, heute ebenfalls verschollenen Deckelpokalen, als Arbeit des 19. Jahrhunderts ein, versehen mit dem Zusatz "Flensburg?"[53] Er dürfte dabei an die in Flensburg tätige Firma E. F. Rönneberg gedacht haben.[54] Deren Gläser sind jedoch hüttentechnisch gearbeitet und haben mit Müllers Produkten keinerlei Ähnlichkeit.[55]

Der Anmerkung Buchers und dem Nürnberger Inventareintrag ist zumindest zu entnehmen, dass in den 80er Jahren in Fachkreisen Informationen über die pseudovenezianischen Gläser Müllers vorlagen und dass man ihre Herstellung in Norddeutschland vermutete. Auch der Pokal I 13a (Abb. 21) wurde von Bucher bereits als Werk des 19. Jahrhunderts eingestuft. Die betreffenden Kenntnisse scheinen jedoch in der Folgezeit wieder verloren gegangen zu sein.[56]

Der Grund für die falsche Einordnung der Gläser C. H. F. Müllers ist in erster Linie darin zu suchen, dass man bis in die heutige Zeit davon ausging, die Stücke seien am Ofen gearbeitet. Dies kann kaum verwundern, sind doch die bis zu 60 cm hohen Gläser nur schwer mit der landläufigen Vorstellung von Lampenarbeit in Einklang zu bringen. Erst bei genauerem Studium wird eine Reihe von Kriterien fassbar, die ihre Identifizierung als lampengeblasene Stücke zweifelsfrei ermöglichen.

Da die Gläser vor der Lampe nur mit Zangen und Greifwerkzeugen bewegt werden, der Vorgang des Umheftens also, von Ausnahmen abgesehen, entfällt, ist der deutlichste Hinweis das Fehlen der Abschlagnarbe.[57] Als weiteres Merkmal kommt hinzu, dass der Fussand stets nach oben umgeschlagen ist, da sich bei der Lampenarbeit der Rand eines Fusstellers beim Auftreiben von sich aus aufstülpt und nur noch angedrückt zu werden braucht. Dagegen wurde aus praktischen Erwägungen und wegen

52. Ein Reflex auf die Publizierung dieses Werkes dürfte in einem Deckelpokal zu sehen sein, der als Teil der Firmensammlung auf S. 10 der um 1922/23 von Salviati herausgegebenen Broschüre *The Glass and Mosaic Industries of Venice* abgebildet ist. Lediglich die Deckelbekrönung wurde variiert. Vgl. auch Ausst. Kat. *Vetri di Murano dell '800*, Museo Vetrario 1978, Kat. Nr. 39, Abb. 67.

53. Vgl. *Die Glassammlung des K. K. Oesterreichischen Museums*, bearbeitet von Bruno Bucher, Wien 1888, S. 83. Wahrscheinlich spielte J. Brinckmann auf diesen Fall an, als er in seinem oben zitierten Erwerbungsbericht für 1893 auf Stücke verwies, die in öffentliche Sammlungen gelangt seien.

54. A. Demmin (s. Anm. 43), S. 61, hatte 1883 in seinem Buch am Rande auf die "nach Venediger Art angefertigten" Gläser dieser Firma verwiesen, die zu viel niedrigeren Preisen als die der Italiener geliefert würden.

55. Zu Rönneberg vgl. Ausst. Kat. Historismus Hamburg (s. Anm. 7), Nr. 339/40, S. 202 f. und S. 484.

56. Der Pokal I 13a z. B. wird in J. Schlossers Handbuch *Das alte Glas*, Braunschweig 1956, Abb. 66, S. 91 wieder als Arbeit des 17. Jahrhundert aufgeführt.

57. Zwar ist es theoretisch möglich, den Mundrand eines Flügelglases auch ohne die Verwendung eines Hefteisens mit Hilfe langer gekröpfter Zangen zu verschmelzen. Wegen des komplizierten Arbeitsablaufs scheint jedoch von dieser Möglichkeit nicht Gebrauch gemacht worden zu sein; zumindest ist mir kein derartiges Glas bekannt geworden.

der besseren optischen Wirkung der Rand bei den weitaus meisten der am Ofen geblasenen venezianischen oder in venezianischer Art gearbeiteten Gläser des 16./17. Jahrhunderts nach unten umgeschlagen.[58] Macht man sich klar, dass ein grösseres Glas vor der Lampe nicht in einem Stück, sondern fortschreitend durch Aneinanderreihung einzelner Blasvorgänge entsteht, so wird deutlich, dass die leicht wellige Kontur grösserer Körper, etwa die Kuppa eines Pokals oder einer Flöte, als weiteres technisch bedingtes Unterscheidungsmerkmal herangezogen werden kann. Auch Unregelmässigkeiten der Deckelkalotte, die sich bei mehreren Gläsern störend auswirken (vgl. Abb. 1, 2, 16, 53, 64), erklären sich aus dem stufenweisen Erhitzen und Ausblasen der Glasröhre.

Weitere Hinweise auf die Entstehungsart eines Glases sind der Gestaltung des Verbindungsstückes zwischen Deckel und Aufsatz zu entnehmen. Abb. 28a zeigt einen schematischen Schnitt durch einen Deckel der Gruppe I. Als Reminiszenz der Entstehung des Deckels aus einer Röhre weist die Kalotte im Scheitelpunkt eine Öffnung auf; dieser Röhrenstumpf wird durch einfaches Umschlagen oder durch Überstülpen und Aufschmelzen eines grösseren Röhrenabschnitts verstärkt und dient so als Auflager für die Bekrönung. Bei hüttentechnisch gearbeiteten Deckeln bleibt die Kalotte geschlossen. Anstelle der Öffnung findet sich eine Abschlagsnarbe, da zum Ansetzen der Bekrönung die Verwendung eines Hefteisens erforderlich ist (Abb. 28b).

Charakteristisch für die lampengeblasenen Gläser ist ferner der gleichbleibende oder nur geringfügig abweichende Durchmesser der Glasstäbe, aus denen Schaftmotiv und Deckelbekrönung gebildet werden. Die Erklärung hierfür liegt in der Verwendung vorgefertigter Fadenglasstäbe—Weichgläser mit hohem Alkaligehalt, die spezialisierte, meist im Thüringer Wald gelegene Hütten als Meterware herstellten. Müller verwendete sowohl fehlerfreies, wie bewusst durch kleine Luftbläschen verunreinigtes Material. Die allzu regelmässige Oberfläche der Fadenglasstäbe suchte er durch zusätzliches Tordieren der erhitzten Stangen zu variieren. Bei vielen Gläsern weist daher die Schaftverzierung eine schnurartig gerippte Oberfläche mit stark gewundenen Spiraleinlagen auf.

Bei den hüttentechnischen Arbeiten des 17. Jahrhunderts bereitete man die Fadengläser für den jeweiligen Zweck meist ad hoc vor; sie wurden nur kurz zwischen den Hefteisen ausgezogen und sind daher oft an den Enden erheblich dicker als in der Mitte.[59] Auch ist die verwendete Glasmasse in der Regel bei Kuppa und Schaft identisch. Bei lampengeblasenen Arbeiten dagegen sind bei diesen Partien häufig Abweichungen im Farbton zu beobachten.

Das Angebot an Fadengläsern scheint zumindest in den Anfangsjahren der Müller'schen Produktion beschränkt gewesen zu sein. Es herrscht die Kombination weiss-rot-gelb-blau vor. Hinzu kommen einfache Farbpaarungen wie weiss-gelb, weiss-rot oder gelb-rot. Bei den in den 60er Jahren entstandenen Gläsern sind die Farbfäden innerhalb der Stäbe voneinander getrennt geführt, bei den späteren Arbeiten überwiegen Stäbe mit bandartig aneinandergefügten Farbstreifen.[60]

---

58. Lediglich bei den schweren Fusschalen der ersten Hälfte des 16. Jahrhunderts überwiegt das Umschlagen nach oben.

59. Häufig wird ein Glasstrang in der Mitte getrennt und die nun konisch zulaufenden Stäbe für zwei Pokale verwendet. Die Schaftfigur wirkt bei diesen Gläsern durch die hierdurch bedingte Asymmetrie besonders lebendig.

60. Die Belieferung der Lampenbläser durch Spezialfirmen macht deutlich, dass Materialeigenheiten, Farbkombination etc. nicht uneingeschränkt für die Zuschreibung von Gläsern an einen bestimmten Künstler herangezogen werden können, da die Produkte der betreffenden Hütten auch jedem anderen zu Gebote standen. Es wird nicht mit vielen derartigen Betrieben zu rechnen sein, da im Thüringer Wald, dem eigentlichen Zentrum der Lampenbläserei in Deutschland, vor dem Ende des 19. Jahrhunderts keine Fadengläser verarbeitet worden zu sein scheinen; vgl. B. Pischel, *Die Thüringische Glasbläserei. Eine volkskundliche Untersuchung über Geschichte und Wesen einer thüringischen*

Weitere Hilfen zur Identifizierung lampengeblasener Gläser können Details wie die Ansätze der Schaftverzierungen bieten. Sollen beispielsweise die Kämme der Schlangen- oder Vogelköpfe nicht mit der Zange gekniffen, sondern im Wortsinne kammartig gegliedert werden, heftet der Glasmacher am Ofen zwei kleine Glasposten an die Schaftschlange, formt sie und schneidet die Musterung mit dem Messer ein. Als Ergebnis dieses Verfahrens zeigt jeweils einer der Kämme die abgeflachte Rückseite, die sich durch das Auflegen des heissen Glases beim Schneiden zwangsläufig ergibt (Abb. 29b). Vor der Lampe dagegen werden ähnliche Motive in der freien Hand aus feinen Glasstäben aufeinandergeschmolzen (Abb. 29a). —Nicht von am Ofen gearbeiteten Stücken zu unterscheiden sind dagegen die mit Hilfe von Zangen gestalteten Partien und die verdickten Überleitungen des Schaftes zu Kuppa und Fuss, obwohl der technische Vorgang hier ebenso von der Hüttentechnik abweicht wie bei den übrigen Arbeitsgängen.[61]

Die Tatsache, dass die Gläser vor der Lampe gearbeitet wurden, sagt zunächst noch nichts über den Zeitpunkt ihrer Entstehung aus. Es wäre daher noch die Frage zu klären, ob es sich bei dem einen oder anderen der Pokale nicht möglicherweise um ein lampengeblasenes Original des 17. Jahrhunderts, also um ein potentielles Vorbild für Müllers Schöpfungen handeln könnte. Hieraus ergibt sich unmittelbar die Frage nach den Grenzen und Möglichkeiten der Lampentechnik in früherer Zeit. Zu ihrer Beantwortung ist ein kurzer Exkurs über Entwicklung und Technologie des Blasens vor der Lampe unerlässlich.

Seit dem Mittelalter, sehr wahrscheinlich sogar bereits in römischer Zeit, bediente man sich für die Herstellung von Glasgerät und Ziergegenständen einer mit Öl oder Talg gefüllten Lampe, deren Flamme durche eine spitz zulaufende Röhre gezielt mit Luft versorgt und so auf die erforderlichen Hitzegrade gebracht wurde. Die Druckluft erzeugte entweder der Lampenbläser selbst oder ein Blasebalg, der mittels eines Tretbrettes betrieben wurde. Ähnliche Geräte dienten seit alters her den Goldschmieden für ihre Löt- und Emaillierarbeiten.

Man geht gemeinhin davon aus, dass Venedig im 17. Jahrhundert auch für den Bereich des lampengeblasenen Glases das entscheidende Zentrum war. Schon eine Nürnberger Handschrift vom Ende des 17. Jahrhundert nennt jedoch als Hauptort für "die Kunst des Glasblasens von allerhand Thierlein, Vögelein, Glässlein, Krüglein und mancherlei Sachen, . . ., von freyer Hand ober einer brennenden Lampen mit Öhl gefüllt" noch vor Venedig die Stadt Mailand.[62] In der Tat spielten die Lampenbläser in Venedig stets nur eine untergeordnete Rolle. Als *perlieri* fertigten sie Glasschmuck und Souvenirartikel, vor allem kleine Tiere und die aus massiven Farbglasstäben geformten Commedia dell'Arte-Figürchen. Angesichts der vollkommenen Hüttentechnik der Muranesen kann die geringe Bedeutung

---

Volkskunst, Weimar 1936, S.45 ff. u. S.76 f. Es wäre sogar denkbar, dass Müller die von ihm verwendeten Fadengläser nach seinen Angaben herstellen liess. Als Lieferanten kämen unter anderen die Glashütte Kühnert & Müller in C. H. F. Müllers Heimatort Piesau und die Firma Greiner Vetter Sohn in Lauscha in Betracht (freundlicher Hinweis Kurt Wallstab, Darmstadt).

61. Der Übergang entsteht bei den Gläsern C. H. F. Müllers folgendermassen: Ausgangsmaterial ist eine Röhre. Diese wird im oberen Teil zur Kuppa ausgeblasen, bzw. zum Fussteller aufgetrieben, in den Schaftpartien zusammengeschmolzen. Die spitze Endung der Kuppa im Schaft, die bei vielen Müller-Gläsern zu beobachten ist (Abb. 28a) und genau mit der bei vielen am Ofen gearbeiteten Stücken übereinstimmt, erklärt sich aus diesem Vorgang. Die Verdickung entsteht durch partielles Erhitzen des Schaftes, der sich beim Hochschieben gegen die Kuppa zusammenstaucht und dabei verbreitert. Anders als bei am Ofen gearbeiteten Stücken sind Schaft und Kuppa also nicht aus zwei verschiedenen Teilen zusammengefügt; dies ist den ohne Verdickung gearbeiteten Füssen (vgl. z.B. Abb. 2) noch ohne weiteres abzulesen.

62. Zitiert nach E.v. Czihak, *Schlesische Gläser*, Breslau 1891, S.98, Anm.1.

dieses Nebenzweiges der Glaskunst in Venedig nicht einmal verwundern.[63]

Die eigentlichen Leistungen der Lampentechnik entstanden im Dienst naturwissenschaftlicher Forschung in Florenz, wo unter der lebhaften Förderung des Grossherzoglichen Hauses der Medici ein Studienzentrum ersten Ranges entstanden war. Bereits 1615–20 setzte hier Galileo Galilei Glas zur Herstellung wissenschaftlicher Instrumente und Messgeräte ein.

Ihren Höhepunkt erreichte die Kunst der florentiner Instrumentenbläser mit den Apparaturen, die sie im Auftrage der Accademia del Cimento, der grossherzoglichen Akademie für naturwissenschaftliche Experimente, zwischen 1657 und 1667 anfertigten.[64] 223 der Instrumente haben sich im Museo di Storia della Scienza in Florenz erhalten, darunter zahlreiche Thermometer, Aerometer und dergleichen.[65] Man wird davon ausgehen können, dass diese mit bewundernswerter Exaktheit gearbeiteten Geräte technisch die Spitze dessen repräsentieren, was im 17. Jahrhundert vor der Lampe geleistet werden konnte. Auffallend ist dabei, dass die Hohlkörper, die bei vielen der Instrumenten unerlässlich sind, einen Durchmesser von ca. 8 cm selten überschreiten, wobei die Wandungsstärke hauchdünn wird.[66] Waren Hohlkörper von grösserem Durchmesser und stabilerer Wandung erforderlich, griff man auf hüttentechnisch gefertigte Gefässe zurück.

Dies bestätigt die Vermutung, dass die Arbeitsmöglichkeiten der ölgespeisten Lampe äusserst beschränkt waren. Zwar konnte man bei entsprechend scharfer Luftzufuhr eine Stichflamme von beträchtlicher Hitze erzeugen, der Flächenwirkung waren jedoch Grenzen gesetzt. Bei grösseren Gegenständen reichte die mit veränderter Luftzufuhr erzeugte breitere Flamme nur noch zum Erwärmen des Werkstückes, nicht aber zur Erreichung der für das Ausblasen erforderlichen Weissglut.

Die Beschränkung der technischen Möglichkeiten spiegelt sich in den schriftlichen Darstellungen des Blasens vor der Lampe, wie sie seit dem 17. Jahrhundert mehrfach veröffentlicht wurden. Den Anfang macht die kleine Abhandlung, die Johann Kunckel bereits der 1679 erschienenen ersten Auflage seiner *Ars Vitraria Experimentalis* am Ende des zweiten Buches *Zur Zugabe* beifügte: "Es ist dieses kleine Glassblasen, so mit der Lampen geschicht, ob zwar nicht der nützlichsten, doch eines der allerzierlichsten Stücke der ganzen Glasskunst; womit gewiss mancherley Ergötzlichkeiten anzustellen." Es folgt eine Beschreibung des Werkplatzes (Abb. 30) und der Arbeitsvorgänge beim Blasen. Die Möglichkeiten des Blasens vor der Lampe beschreibt er so: "wer nun hier das Zeichnen und Posiren wohl versteht, der kann auf diese Weise in das Glass, was er nur will, posiren, als Bilder, Figuren, Cruzifix, kleine Gefässe etc. ja was man nur erdenken kann; wozu man denn auch kleine Zänglein, und von Draht zusammengebogene Klüfftlein haben muss." Es wird deutlich, dass die vielfältigen Möglichkeiten der Gestaltung in Kunckels Augen ausschliesslich auf das kleine Format beschränkt sind. In erster Linie waren

---

63. Aufschlussreich ist ein Brief des venezianischen Gelehrten Giovan Francesco Sagredo an Galileo. Er schreibt 1618: "Hier [in Venedig] haben wir in dieser so schönen Kunst keine Leute, die etwas taugen und wenn es sich ergibt, dass ich mir irgendetwas anfertigen lassen muss, so muss ich die Arbeit beaufsichtigen und denen, die sie machen, vieles erst noch beibringen." Den Hinweis auf diesen Brief und zahlreiche erhellende Informationen zur Bedeutung der Lampenbläserei in Venedig und Florenz verdanke ich Herrn Astone Gasparetto, Venedig.

64. Vgl. M. Celeste Cantie u. M. Luisa Righini Bonelli, *The Accademia del Cimento*, o.O., o.J.

65. Die Instrumente sind publiziert im Anhang der Faksimile Ausgabe der 1666/67 erschienenen Publikation von Experimenten der Accademia, in der zahlreiche weitere Geräte in Kupferstichen abgebildet sind; vgl. *I Saggi di naturali esperienze fatte nell'Accademia del Cimento e Strumenti e Suppellettili della medesima Accademia Conservati presso il Museo di Storia della Scienza di Firenze* . . ., Pisa 1957.

66. Zu den Stücken mit dem grössten Durchmesser gehören die Spiralthermometer, von denen Exemplare erhalten sind, deren Flüssigkeitsbehälter einen Durchmesser von 6, 8 cm aufweist; vgl. Cantie/Righini Bonelli (s. Anm. 64), S. 10 mit Abb.

es die—von ihm nicht erwähnten—Perlen, die ihre Entstehung der Lampenarbeit verdankten, dann die im 17. und 18. Jahrhundert sehr beliebten, vor allem in Italien, den Niederlanden und Frankreich hergestellten Glasfigürchen und aufs feinste gearbeiteten Krippenszenerien, Kreuzigungsdarstellungen etc.[67] Die erwähnten kleinen Gefässe fanden ihren Platz in den ebenfalls sehr geschätzten und verbreiteten Puppenhäusern, die das tägliche Leben ihrer Zeit wiederspiegeln und von denen sich vornehmlich in den Niederlanden und in Mitteldeutschland besonders reich ausgestattete Exemplare erhalten haben.[68]

Den geringen Wandel in der technischen Ausstattung einer Lampenbläserwerkstatt vom 17. zum 18. Jahrhundert veranschaulicht ein Kupferstich aus einem 1769 in Wien erschienenen Traktat (Abb. 31).[69] Die Gerätschaften sind im Wesentlichen unverändert. Gezeigt wird ein Glasspinner bei der Arbeit, ergänzt durch eine Auswahl der Objekte, die im 18. Jahrhundert vor der Lampe geblasen wurden und auf deren Herstellung in dem Traktat näher eingegangen wird. Aus dem Bereich künstlerischer Glasgestaltung sind dies Glasblumen, Appliken für einen Tafelaufsatz, ein Schwan oder eine Gans, und ein *Cartesianischer Teufel*, eines jener beliebten kleinen Glasfigürchen, mit deren Hilfe sich die physikalischen Gesetze des Luft- und Wasserdrucks in spielerischer Weise demonstrieren liessen.[70]

Der Zusammenstellung des Wiener Traktats entspricht die Aufzählung von Lampenarbeiten durch Georg Ludewig Hochgesang. In seiner 1780 veröffentlichen *Historischen Nachricht* erwähnt er neben den "Cartesianischen Männerchen" noch "Springerbsen, Schreibfedern, Wasserhammer, Bier- Branntwein, und Weinproben, Wasserwaagen, Korallen und dergleichen mehr."[71]

Der Arbeit an der Verbesserung der Lampe gilt, beginnend mit der 1788 erschienenen *Oekonomisch-technologischen Encyclopädie* des Johann Georg Krünitz,[72] der die Anreicherung der Druckluft mit Weingeistdämpfen propagiert, das Interesse der Verfasser fast jeder der zahlreichen späteren Lehrschriften. Besonders Versuche mit verschiedenen Brennstoffen, darunter auch Palmöl, Kokosfett, Petroleum und Benzin, sowie die Verbesserung der Luftzufuhr spielen dabei eine Rolle.[73] Keiner der sehr detaillierten Darstellungen aus der ersten Hälfte des 19. Jahrhunderts sind irgendwelche Hinweise auf Trinkgefässe oder andere Arbeiten grösseren Formats in der Art der Müller'schen Flügelgläser zu entnehmen. Dies gilt für die 1829 publizierte grundlegende

---

67. Zentrum der Lampenbläserei in Frankreich war Nevers, in den Niederlanden Amsterdam; vgl. die Reisebeschreibung des Filips von Zesen aus dem Jahre 1664, zitiert bei F. W. Hudig, *Das Glas*, Diss. Zürich, Wien 1923, S.51 f.

68. Für die Niederlande vgl. B. Jansen, Glassware in Dutch Dolls Houses, in: *Annales du 2d Congrès des "Journées Internationales du Verre,"* Leiden 1962, S.117 ff. Für Mitteldeutschland vgl. Wolfgang Leber, *Die Puppenstadt Mon Plaisir*, Museen der Stadt Arnstadt 1975, Abb. S.27, 47.

69. J. d. B., *Die Glasschmelzkunst bey der Lampe oder Anweisung, wie die Glasröhren zu schmelzen und daraus allerley Figuren zu blasen sind . . .*, Wien 1769.

70. Das hohlgeblasene Teufelchen konnte z.B., frei in einem mit Wasser gefüllten Gefäss schwebend, durch Ansteigen oder Absinken die Veränderung des Luftdrucks anzeigen und damit Wetteränderungen ankündigen.

71. Vgl. G. L. Hochgesang, *Historische Nachricht von Verfertigung des Glases*, Gotha 1780, S.65.

72. J. G. Krünitz, *Oeconomisch-technologische Encyclopädie oder allgemeines System der Stats-Stadt-Haus- und Land-Wirthschaft, und der Kunst-Geschichte in alphabetischer Ordnung, Achtzehnter Theil, von Ger bis Glasur, Zweyte Auflage*, Berlin 1788, S.650–657 mit Fig. 980a, b. — Die Encyclopädie von Diderot und d'Alembert geht unter dem Stichwort *Emailleur* nur am Rande auf die Technik der Lampenarbeit ein; vgl. Bd. XXVI, Arts et Metiers Mechanique II, Coul-Filets, Bd. XXXVI, Arts et Metiers Mechanique X, Planches 2, Pl. 1 u.2, fol.30. Unter dem Stichwort *Verre* finden sich nur Herstellungshinweise für wissenschaftliche Instrumente; vgl. Bd.XXXII, Arts et Metiers Mechanique VII, Tab-Zim, Paris 1771, S.562–65.

73. Es war nicht möglich, die bessere Leistung der Lampe durch einfache Vergrösserung von Ölgefass und Docht zu erreichen, weil die Russentwicklung ab einer bestimmten Dochtstärke so sehr zunahm, dass sie die saubere Bearbeitung des Glases unmöglich machte. Das Russen der Flamme war das Hauptproblem bei der Arbeit mit pflanzlichen oder tierischen Brennstoffen.

Arbeit F.P. Dangers[74] ebenso wie für die Anleitungen Friedrich Körners, Heinrich Rockstrohs und des Franzosen Lafond;[75] auch die umfassenden Darstellungen in Heinrich Lengs Handbuch und der von Joh. Jos. Prechtel herausgegebenen *Technologischen Encyclopädie*,[76] ja selbst die das Wissen der Zeit zusammenfassende Gesamtdarstellung Emanuel Schreibers von 1849 ergeben nichts.[77] Zieht man als Ergänzung die erhaltenen Auftragsbücher des 19. Jahrhunderts von Lampenbläsern des Thüringer Waldes heran, die neben Thermometern und dergleichen ebenfalls ausschliesslich kleinformatige Spielzeugmöbel, Glasfrüchte, Figurengruppen, Tiere, Christbaumschmuck, Hutnadeln usw. aufführen,[78] so wird man zu dem Schluss kommen müssen, dass bis zur Mitte des 19. Jahrhunderts wegen der technischen Unzulänglichkeiten der zum Lampenblasen verwendeten Geräte und Materialien die Arbeit in dieser Technik an das kleine Format gebunden war.

Dies änderte sich nachhaltig mit der Einführung leuchtgasbetriebener Brenner anstelle der alten Öllampen. Die Umstellung auf die neue Technik, die sich in den ersten beiden Jahrzehnten der zweiten Jahrhunderthälfte vollzog, löste schlagartig alle Probleme.[79] Die höheren Hitzegrade und die grössere Arbeitsflamme gaben dem Lampenbläser nun die Mittel an die Hand, sogar mit den Arbeiten der Hütten in Wettbewerb zu treten. Die Gläser C.H.F. Müllers sind hierfür ein sprechender Beleg. Führt man den Gedanken zu Ende, so ergibt sich für die Einordnung und Datierung grossformatiger, dickwandiger Gläser ein technisch bedingter Terminus: wenn sie zweifelsfrei vor der Lampe geblasen sind, können derartige Stücke nicht vor der Mitte des 19. Jahrhunderts entstanden sein.

Der Verzicht auf die Herstellung von Nachschöpfungen und Imitationen venezianischer Gläser seit der zweiten Hälfte der 70er Jahre brachte für Müllers kunstgewerbliche Produktion eine entscheidende Zäsur, bedeutete aber keinesfalls die völlige Aufgabe dieses Zweiges seiner Tätigkeit. In der Hamburger Sammlung hat sich eine 1879 erworbene Flöte mit geripptem Balusterschaft und Fuss erhalten, die laut Inventar "nach Zeichnung und Angabe von F. Keller-Leuzinger von Glasbläser Müller in Hamburg" angefertigt wurde (II 1; Abb. 32/1). Das Glas unterscheidet sich so grundlegend von den bisher gezeigten, dass eine Zuschreibung an Müller ohne den Beleg des

---

74. Dangers Buch wurde bereits 1831 ins Englische übersetzt und erschien 1833 in Quedlinburg und Leipzig unter dem Titel *Die Kunst der Glasbläserei vor dem Löthrohre und an der Lampe* auch in deutscher Sprache.

75. Fr. Körner, *Anleitung zur Bearbeitung des Glases an der Lampe*, Jena 1831 (mit ausführlicher Erörterung der Vor- und Nachteile der verschiedenen Brennmaterialien); H. Rockstroh, *Die Glasblasekunst im Kleinen*, Lissa und Leipzig 1833; Ueber die Kunst Glas zu blasen, von Hrn. Lafond, mit Verbesserungen des Hrn. Danger, *Polytechnisches Journal*, hrsg. von Dr. Johann Gottfried Dingler, Jg. 1933, Nr. XVIII, S.121–140.

76. *Neuer Schauplatz der Künste und Handwerke*, 79. Band, H. Leng, *Vollständiges Handbuch der Glasfabrikation nach allen ihren Nebenzweigen . . .*, Weimar und Ilmenau 1835, S.500–520. *Technologische Encyclopädie oder alphabetisches Handbuch der Technologie, der technischen Chemie und des Maschinenwesens . . .*, hrsg. von Joh. Jos. Precht, 7. Band, Glasblasen — Hutmacherkunst, Stuttgart 1836, S.1–18.

77. *Neuer Schauplatz der Künste und Handwerke*. Mit Berücksichtigung der neuesten Erfindungen. 177. Band, E. Schreiber, *Die Glasblasekunst sowohl auf der Glashütte, als an der Glasbläserlampe . . .*, Weimar 1849, S.36–185.

78. Vgl. Ernst Löber, *Zur Geschichte der Lampenbläserei auf dem Thüringer Walde*, Weimar 1926, S.13 ff. und B. Pischel (s.Anm.60), S.46 ff.

79. Vergleicht man die Arbeiten in zwei im Glasmuseum des thüringischen Glasbläserzentrums Lauscha einander gegenübergestellten Vitrinen—auf der einen Seite die kleinen, in herkömmlicher Technik gefertigten Figürchen, auf der anderen die erheblich grösseren, vor dem Gasgebläse entstandenen Stücke—so wird die Bedeutung der neuen Entwicklung unmittelbar anschaulich. Vgl. Rudi Hofmann, *Das Museum für Glaskunst Lauscha. Ein Spiegelbild volkskundlichen Schaffens*, Sonneberg 1971, S.23, Abb. S.24/25; für Lauscha lässt sich der Zeitpunkt der Umstellung auf die neue Technologie mit dem Jahr der Errichtung der ersten Gasanstalt 1867 genau bestimmen. — In Venedig, wo im 19. Jahrhundert fast ausschlieslich Perlen vor der Lampe gefertigt wurden, setzte sich, nachdem der entsprechende Brenner bereits 1843 erfunden worden war, die Arbeit mit der Gasflamme zwischen 1850 und 1855 allgemein durch, vgl. B. Cecchetti/V. Zanetti, *Monografia della vetrariâ veneziana e muranese*, Venedig 1874, S.136 ff.

Inventareintrags wohl kaum in Betracht gezogen worden wäre. Die hornfarbene Glasmasse und die Verwendung einer gerippten Röhre für Fuss und Schaft des Glases sind ebenso neu im Werk Müllers, wie die aufgelegten farbigen Flach- und Perlnuppen. Selbst Details wie der Fussrand, der nicht mehr umgeschlagen, sondern als massiver Wulst gestaltet ist, weichen vom gewohnten Bild ab. Bei einem weiteren Glas dieser Art (II 1b) bestand der Kelch sogar aus rötlichem Glas.

Angesichts der Inventarnotiz erinnert man sich an zeitgenössische Aussagen, denenzufolge Müller "sowohl nach eigener Erfindung, wie nach Modellen älterer Gefässe oder nach den Zeichnungen bewährter Künstler" gearbeitet hat, wobei es sein Bestreben war, "die Formen und die Technik des venezianischen Glases wieder dem täglichen Gebrauch dienstbar zu machen."[80] Offenbar hatte die Einsicht in die Grenzen des eigenen Gestaltungsvermögens ihn zu dem 1877–79 in Hamburg tätigen Maler, Schriftsteller und Kunstgewerbeentwerfer Franz Keller-Leuzinger geführt,[81] der für ihn eine heute nicht mehr genau bestimmbare Anzahl von Formen entwickelte. Genannt werden im Hamburger Inventar neben der Flöte ein "Spitzglas," ein "Bierglas mit Warzen." Ein "ovalgedrücktes Glas," ein "Römer" und ein "Weinglas," wurden 1882 an Heinrich Sauermann, den Leiter des Flensburger Gewerbemuseums verkauft und konnten im Bestand des dortigen Städtischen Museums aufgefunden werden (II 4, 3, 2; Abb. 32/2).[82] Die Gläser lehnen sich an Glasformen des Nordens an, beziehen sich also nicht auf die venezianische Tradition. Anders als bei den übrigen Stücken der Gruppe sind die Füsse nicht aus gerippten, achsial tordierten Röhren gefertigt; sie wurden vielmehr aus glatten Röhren aufgetrieben und zusätzlich mit einem Faden umwickelt. Die drei Gläser belegen den oben zitierten Hinweis Brinckmanns, demzufolge auch "alte deutsche" gläser aus Müllers Werkstatt hervorgegangen sind. Formal schliesst die Gruppe relativ eng an deutsche Trinkgläser des 16./17. Jahrhunderts an; der Besatz mit andersfarbigen Nuppen und Fäden jedoch folgt den dekorativen Vorstellungen des Historismus.

Den Gläsern nach Entwurf Keller-Leuzingers kann ein Römer im Stockholmer Nationalmuseum angeschlossen werden (II 5; Abb. 33), dessen Nuppendekor dem der Flöte II 1 (Abb. 32/1) nahesteht. In der allgemeinen Auffassung wie im Detail ist dieser Flöte ferner ein Kelchglas sehr ähnlich, das ein Jahr später in die Sammlung des Hamburger Museums gelangte (II 6; Abb. 34/1). Erstaunlicherweise ist als Provenienz die Kunstgewerbliche Werkstatt R. Bichweiler in Hamburg angegeben, mit dem ausdrücklichen Vermerk "nach Zeichnung von R. Bichweiler," der Name Müller ist nicht genannt. Heinrich Robert Bichweiler, Architekt, Bildhauer und rühriger Kunstgewerbeentwerfer, widmete sich als Leiter der 1878 gegründeten "Kunstgewerblichen Werkstatt" vor allem der Wiederbelebung keramischer Techniken. Daneben galt sein Interesse in erster Linie der Kunsttischlerei und der Metalltreibarbeit.[83] Das Glas spielte daneben nur eine untergeordnete Rolle. Da die von seiner Firma angebotenen vielfältigen Artikel nur zum Teil in eigenen Werkstätten gefertigt wurden, wird man davon ausgehen können, dass Bichweiler die von ihm entworfenen Gläser durch C. H. F. Müller ausführen liess. Anders ist die enge Verwandtschaft der Flöte und des Kelchglases nicht zu erklären, es sei denn, man geht von der Annahme aus, Bichweiler habe die ent-

80. Kat. III. Deutsche Kochkunstausstellung, *Glas*, Hamburg 1880, S.7/8; vgl. auch den Vermerk im Ehrenbuch der Röntgenologen (s.Anm.22), S.20: "Diente ihm anfänglich Venedig mit seinen Gläsern als Muster, so hat er später vollkommen frei und grosszügig geschaffen."

81. Zu Fr. Keller-Leuzinger vgl. Thieme/Becker, *Allgemeines Lexikon der bildenden Künste*, Bd.XX, Leipzig 1927, S.116 und Ausst. Kat. Historismus Hamburg (s.Anm.7) S.465 f.

82. Für die Übermittlung von Angaben zu den drei Gläsern habe ich Herrn Ulrich Schulte-Wülwer zu danken.

83. Zu R. Bichweiler vgl. Ausst. Kat. Historismus Hamburg (s.Anm.7), S.439 ff.

sprechende Fachkraft bei Müller abgeworben. Wahrscheinlicher ist jedoch, dass Müller, der sich zunehmend auf die Produktion Geissler'scher Röhren und technischer Geräte zu konzentrieren begann, im Auftrag der Kunstgewerblichen Werkstatt arbeitete. Da Müller über kein eigenes Ladengeschäft verfügte und dies Verfahren ihn von der Notwendigkeit befreite, sich selbst um den Absatz der Gläser kümmern zu müssen, war es auch für ihn von Vorteil.

Die enge Verwandtschaft der gestalterischen Vorstellungen Bichweilers und Keller-Leuzingers belegen zwei Flöten in Eisenach (II 7a) und Corning (II 7b; Abb. 34/2), bei denen die Entscheidung für die Autorschaft des einen oder des anderen unmöglich ist. Nachweislich auf Bichweiler gehen ausser II 6 die Entwürfe für zwei Weingläser im Hamburger Museum (II 8 a, b; Abb. 35), einen Spitzkelch im Nationalmuseum Stockholm (II 9; Abb. 37) und eine Flöte im Kunstmuseum Düsseldorf zurück (II 10; Abb. 38), die teils unmittelbar von seiner Firma bezogen wurden, teils durch die Beischrift einer Skizze Oskar Rauters (Abb. 36) als Arbeiten Bichweilers gesichert sind.[84] Allen gemeinsam sind zwei auffällige Motive—grosse, von gesondert ausgeblasenen Kügelchen gesäumte Hohlscheiben und kleine, sehr regelmässige Schaftringe oberhalb des Fusses und unterhalb der Kuppa. Beide Details sind kennzeichnend für lampengeblasene Arbeiten und am Ofen in gleicher Weise nicht oder nur unter grossen Schwierigkeiten herzustellen.[85]

Diesen Gläsern ist ein aufwendiger Pokal anzuschliessen, dessen Schaft dem der Flöte II 10 entspricht (II 11; Abb. 39). Neu ist die Gestaltung der Kuppa, die aus zwei miteinander verschmolzenen glatten und gerippten Röhrenabschnitten geblasen wurde, auch dies ein Verfahren, das nur vor der Lampe möglich ist. —Der gemeinsam mit II 6 und 8 in die Hamburger Sammlung gelangte Senftopf mit Löffel (II 12; Abb. 40) mag Bichweilers Bemühen um die kunsthandwerkliche Gestaltung selbst einfachen Gebrauchsgeräts beispielhaft veranschaulichen.

In den Grenzbereich zwischen Gebrauchsglas und Zierglas gehört eine Gruppe von zum Teil überdimensionalen Balusterflöten, von denen sich Exemplare in zahlreichen Museen und Sammlungen nachweisen lassen (II 13; Abb. 41a, b). Ihrer ungewöhnlichen Grösse wegen—sie messen bis zu 60 cm—wurden sie bisher allgemein als am Ofen entstandene Arbeiten des 17./18. Jahrhunderts eingeordnet. Die Glasmasse, die Verwendung einer gerippten Röhre für Fuss und Schaft und einer glatten für den Kelch sowie die beiden Schaftringe belegen jedoch hinreichend die Zugehörigkeit zum Kreis um Müller, Keller-Leuzinger und Bichweiler.

Die Zusammenarbeit mit den engagierten Vertretern fortschrittlichen Kunsthandwerks führte Müller zu einem völlig neuen Verständnis seiner Technik und seines Materials. Die Gläser der Gruppe II sind völlig frei von dem Bemühen, einem in der Hütte gearbeiteten älteren Vorbild nahezukommen, und die Lampentechnik ist nicht länger unzureichender Ersatz für die nicht verfügbare, aufwendigere Hüttentechnik. Die neuen Formen sind vielmehr konsequent aus den Möglichkeiten der Lampentechnik entwickelt und nutzen bewusst deren Vorteile. Sie erfüllen damit eine der Hauptforderungen der Kunsttheoretiker des Historismus, die der materialgerechten Ar-

---

84. Die Beischrift lautet: *Geschenk von Tacchi (fabr. v. Bischweiler Hamburg)*. Die Firma P. A. Tacchi & Co in Frankfurt war ein bedeutendes Glasverlagsunternehmen; vgl. Kat. Historismus Berlin (s. Anm. 5), Anhang. — Der Hinweis auf Tacchi ist wichtig, da er belegt, dass die Müller/Bichweiler'schen Gläser auch ausserhalb Hamburgs vertrieben wurden.

85. Die Kügelchen am Rand der aus einer gestauchten Kugel entwickelten Hohlscheibe bilden sich nach punktuellen Erhitzen bei gleichzeitigem Ausblasen. Um das Durchbrechen des Luftstroms zu verhindern, ist es dabei unerlässlich, die erhitzte Partie durch Aufschmelzen eines kleinen Glaspostens zu verstärken; die Schaftringe entstehen durch einfaches Stauchen des an der betreffenden Stelle in Weissglut gebrachten Glases.

beitsweise, die den durch die Technik gesetzten Bedingungen und Eigenheiten Rechnung trägt. Es entstanden so im besten Sinne historistische Gläser, die, obwohl der venezianischen Glaskunst verpflichtet, in Form und Detail eindeutig der zweiten Hälfte des 19. Jahrhunderts angehören.

Es versteht sich, dass der Zuwachs an Substanz, den Müllers Gläser durch die Entwürfe Keller-Leuzingers und Bichweilers gewonnen haben, nicht ihm gutgeschrieben werden kann; sein Anteil ist der des technisch perfekten ausführenden Handwerkers. Inwieweit er selbst die Erkenntnisse, die ihm der Kontakt mit den beiden Fachleuten vermittelte, bei einer dritten Gruppe von Gläsern zu verwerten vermochte, in der die neuen Arbeitsprinzipien zum Teil auf die alten Gefässtypen der Gruppe I übertragen wurden, ist nicht mit Sicherheit zu sagen. Die Mehrzahl der Gläser dieser Gruppe zeugt von so überragender Meisterschaft, dass man geneigt ist, für den Entwurf auch hier die Hand eines gelernten Kunstgewerbezeichners vorauszusetzen.

Dies gilt sicher noch nicht für den am Anfang der Reihe stehenden Deckelpokal, der 1915 als Arbeit Müllers aus Privatbesitz in die Hamburger Sammlung gelangte (III 1; Abb. 42). In mancher Hinsicht erfüllt er eine Mittlerfunktion zwischen den Gläsern der Gruppe I und den Pokalen der Gruppe III. Das verwendete Material ist bereits das bräunlich hornfarbene Glas der Flöte II 1a. Auch wurde jede Nachahmung hüttentechnisch bedingter Details vermieden. Die Form der Kuppa ist noch stark den älteren Typen verhaftet, das Deckelprofil weist dagegen bereits jene geschwungene Kontur auf, die, leicht variiert und verfeinert, für alle Pokale der Gruppe III verbindlich werden sollte. Auch der Übergang von der Deckelkalotte zur Bekrönung wurde ganz im Sinne überlegt angewandter Lampentechnik vereinfacht und klarer gelöst (vgl. Abb. 28c).[86] Dennoch bleibt der Gesamteindruck, nicht nur der verlorenen Kämme wegen, etwas unbefriedigend. Man wird davon ausgehen können, dass der Entwurf von Müller selbst unter dem Eindruck der Vorstellungen Keller-Leuzingers und Bichweilers entwickelt wurde. Die Ausführung scheint Werkstattkräften überlassen worden zu sein, die handwerklich noch nicht ganz perfekt waren.

Ein völlig anderes Bild ergibt sich angesichts eines Pokals in Corning (III 2; Abb. 43), der, obwohl in gleicher Weise beschädigt, den Eindruck vollendeter technischer Ausführung nach einem vorzüglichen Künstlerentwurf vermittelt. Den Ausgangspunkt bildete der Typ des Doppelacht-Pokals (vgl. Abb. 1, 2), der hier jedoch—auf der Grundlage der technischen Gegebenheiten der Lampentechnik—neu durchdacht und gestaltet wurde. Fuss und Schaft sind in der von den Gläsern der Gruppe II her geläufigen Weise aus einem gerippten Röhrchen geblasen, die Kuppa ist in der Art des Glases II 8 (Abb. 39) aus unterschiedlichen Rohrabschnitten verschmolzen. Auch ihre schlanke, spitz zulaufende Form entspricht diesem Stück. Auffallend ist die organischere Führung der Stäbe des Schafteinsatzes, die zwar den abgetreppten Umriss der älteren Doppelacht-Pokale beibehalten, deren etwas ungefüge Sperrigkeit jedoch vermeiden.

Dem gleichen Prinzipien folgt die Entwicklung des Frankfurter Pokals III 3a (Abb. 44) aus dem Typ I 5 (vgl. Abb. 8). Die hohe Perfektion der Ausführung—ein Hauptanliegen historistischer Glaskunst—kontrastiert bei diesem Pokal besonders deutlich mit der ungelenken Gestaltung des früheren Stückes; dessen Zufälligkeiten und Abweichungen von der Symmetrie sollten es offensichtlich dem lebendigen Erscheinungsbild in der Hütte gearbeiteter alter Gläser näherbringen. Eine zweite, geringfügig abweichende Version des

---

86. Schmilzt man eine Glasröhre an einem Ende zusammen und bläst sie unter der Schmelzstelle auf, so entsteht zwangsläufig die Form des Anschlusses, die die Deckel der Gruppe III aufweisen.

Glases in Frankfurt hat sich im Bremer Focke Museum erhalten (III 3b).

Es liegt auf der Hand, dass die Spiralschlangen-Pokale des Typs I 9/10 durch die Anpassung an das neue Gestaltungsschema besonders gewinnen mussten. Zwei Pokale in Stockholm folgen teils genau dem älteren Vorbild (III 4a), teils variieren sie es leicht durch den Verzicht auf die Schnabelfortsätze und die Vereinfachung der Kämme (III 4b; Abb. 45). Die Version mit angesetztem Drachenkopf und -schwanz lässt sich ebenfalls in zwei Exemplaren nachweisen. Das des Düsseldorfer Kunstmuseums (III 5a; Abb. 46) konnte, einer Anregung Axel von Salderns folgend,[87] kürzlich durch den im Musée du Verre in Lüttich erhaltenen Deckel komplettiert werden.[88] Das vollendetste Glas der Gruppe III ist ohne Zweifel das der Sammlung in Corning (III 5b; Abb. 47); es besticht durch die Spannung des Aufbaus bei völliger Harmonie in der Abfolge der heterogenen Einzelformen. Auch bei diesen Gläsern führten Vereinfachungen im Detail zur Steigerung der Gesamtwirkung: die Drachenköpfe sind massiv belassen und vermeiden so die etwas plump-unförmige Wirkung ihrer hohlgeblasenen älteren Vorläufer. Dem Stück in Corning kann der phantasievolle Schlangenpokal in Stockholm (III 7; Abb. 49) als qualitativ gleichrangig an die Seite gestellt werden; auch er geht formal auf frühere Arbeiten Müllers zurück (vgl. Abb. 53, Nr. 20).

Neben freien Varianten der erprobten Typen (III 6; Abb. 48) wandten sich die Künstler, für die Müller tätig war, auch der Entwicklung völlig neuer Formen zu, für die sich in der Glaskunst des 17. Jahrhunderts keine unmittelbaren Vorbilder finden lassen. Ein Hervorragendes Beispiel hierfür ist das Zierglas des Frankfurter Museums (III 8; Abb. 50), dessen nuppenbesetzte, aus Fadenglasstäben gebildete Schaft- und Deckelkugel in dieser Form nur im 19. Jahrhundert möglich war.

Nur entfernt an Gläser wie die Blütenpokale I 12 (vgl. Abb. 20) erinnern zwei deckellose Ziergläser in Stockholm (III 9a, b; Abb. 51), bei denen ein weiteres Mal Details aus der Gruppe der Gebrauchsgläser aufgenommen wurden (vgl. die Schäfte Abb. 35, 37). Das gleiche Motiv bestimmt die Gläser des letzten derzeit nachweisbaren Typs, die ihren besonderen Reiz einem höchst dekorativ um die Schäfte gewundenen Schlangenpaar verdanken (III 10c, g, f; Abb. 52a–c). Bei diesen Stücken wird der Zusammenhang mit C. H. F. Müller auch dokumentarisch fassbar. Unter den 1893 durch Justus Brinckmann erworbenen Gläsern befinden sich zwei, um deren "balusterförmigen Stengel sich zwei symmetrisch angebrachte Drachen" wanden. Auch die farbigen Spiralfäden im Innern der Drachen und die blaugrünen Ansätze sind erwähnt (III 10a, b). Der Variationsspielraum bei der Verzierung der Gläser dieses Typs war besonders gross. Er reichte von der differenzierten Gestaltung der Kuppa durch in Wellen oder streng vertikal geführten Rippendekor (III 10c, g; Abb. 52a, b) oder reliefartig aufgeschmolzene Fäden (III 10f; Abb. 52c) bis zur Verwendung einfarbig blauer oder feingeriefelter farbloser Glasstäbe für die Schlangenleiber (III 10a, f).

Der unbestreitbar hohe Rang der Gläser der Gruppe III, deren handwerkliche Vollendung bei einzelnen Stücken kaum zu übertreffen ist, macht ein weiteres Mal deutlich, wo die Stärken C. H. F. Müllers und seiner Werkstatt lagen. Die perfekte Beherrschung von Material und Technik, die noch den Kunstglasbläsern der Gegenwart uneingeschränkte Hochachtung abfordert, spiegelt die Erfahrungen bei der Fertigung der einfacheren Typen der 60er und 70er Jahre. Dass die eigenständige Erfindungskraft demgegenüber zurücktritt und Müller stets an die Auseinan-

---

87. Vgl. v. Saldern (s.Anm.37), S.68.

88. Der Deckel wurde dem Kunstmuseum Düsseldorf dank des freundlichen Entgegenkommens des Leiters des Musée du Verre, Herrn Joseph Philippe, als Leihgabe sur Verfügung gestellt.

dersetzung mit Vorbildern gebunden bleibt, darf nicht verwundern. Die Tradition der künstlerischen Glasbläserei des Thüringer Waldes, der er entstammte, fand ihren Ausdruck im Bewahren, Abwandeln und Weitergeben vorgeprägter Formen und Typen.[89] Gläser leicht abgewandelt nach Vorbildern zu blasen, kam also Müllers gewohnter Arbeitsweise entgegen. Die Initiative zur Wiederaufnahme altvenezianischer Formen ging aber sicher nicht von ihm aus—im Thüringer Wald kam es erst in den 80er Jahren des 19. Jahrhunderts zu ersten Auseinandersetzungen mit der venezianischen Glaskunst[90]—sondern von den durch Brinckmann erwähnten fremden Antiquitätenhändlern.

Festzuhalten ist der frühe Zeitpunkt des Rückbezuges auf Venedig. Die besagten Händler hatten äusserst schnell auf das sich lebhaft regende Interesse der Zeit an den venezianischen Gläsern reagiert, ebenso wie dies zur gleichen Zeit Salviati mit seinen intensiven Bemühungen um die Erneuerung der Glaskunst in Murano getan hatte.[91]

Fragt man nach dem Stellenwert vor allem der späteren Gläser aus Müllers Werkstatt innerhalb der historistischen Glaskunst Deutschlands, so wird man im Bereich des unveredelten frei geblasenen Glases kaum etwas finden, das den Stücken der Gruppe III an Delikatesse der Formgebung und Feinheit der Ausführung ebenbürtig wäre. Weder in Theresiental oder in der Josephinenhütte noch bei Heckert wurden Stücke vergleichbarer Qualität gefertigt. Die Gläser C. H. F. Müllers gehören damit zur absoluten Spitze der deutschen Glaskunst des Historismus, eine Stellung, die ihnen erst in den 80er Jahren mit der Gründung der Abteilung für Kunst-Erzeugnisse der Köln-Ehrenfelder Hütte streitig gemacht werden sollte.

Nach Abschluss meines Manuskripts wurden durch Axel von Saldern im Hamburger Museum für Kunst und Gewerbe zwei Fotografien aus dem vergangenen Jahrhundert aufgefunden (Abb. 53, 54), für deren Überlassung ich ihm an dieser Stelle danken möchte. Aufgrund der Identität des Glases Nr. 21 (Abb. 53) mit dem durch Brinckmann erworbenen Glas I 1g (Abb. 1)[92] lassen sie sich eindeutig als Musterblätter der Werkstatt C. H. F. Müller bestimmen. Die Aufnahmen zeigen eine Auswahl von Gläsern, die in den 70er Jahren zu Müllers Angebot gehörten, im wesentlichen Exemplare der in Gruppe I zusammengefassten Typen (vgl. Abb. 3, 4, 6, 7, 13, 14 ff., 19, 21, 26, 27). Einige der Gläser lassen sich vorerst nur auf den Musterfotos nachweisen (Nr. 3, 5, 6, 9, 12, 15, 19, 20).[93]

Bei näherer Betrachtung bestätigt sich der Eindruck, dass bei den in Müllers Werkstatt

89. Dies macht ein Blick in die bei verschiedenen Artikeln über Jahrzehnte gleichbleibenden Einträge in den erhaltenen Auftragsbüchern deutlich (s. Anm. 78). Vgl. auch Pischel (s. Anm. 60), S. 54 f.

90. Die entscheidende Rolle spielte hier Friedrich Zitzmann, der zu dieser Zeit noch in Lauscha ansässig war, und bereits seit 1885 venezianisierende Gläser en führende europäische Museen lieferte.

91. Das allgemeine Interesse an der Glaskunst Venedigs belegt sehr anschaulich der Preis, den die Londoner Society of Arts 1864 für die möglichst exakte Kopie eines Flügelglases aus dem Bestand des Victoria & Albert Museums aussetzte. Die Beteiligung allerdings blieb gering. Eine äusserst zufriedenstellende von Marco Seguso geblasene Nachbildung lieferte die Firma Salviati; sie kam jedoch für die Zuerkennung eines Preises nicht in Betracht, da sie nicht in England gefertigt worden war (vgl. *Journal of the Society of Arts*, 1. Februar, 1867, S. 161). Der einzige vergebene Preis wurde schliesslich Joseph Leicester zuerkannt. Der Mitteilung ist nicht zu entnehmen, ob Leicester vor der Lampe oder am Ofen arbeitete; vgl. *Journal* . . . , 14. Februar, 1868, S. 241. Den Hinweis auf diesen Vorgang verdanke ich Herrn J. V. G. Mallet, Victoria & Albert Museum, London.

92. Der Detailvergleich belegt die Identität zweifelsfrei. Zu beachten ist, dass der Pokal auf Abb. 53 gegenüber Abb. 1 von der Rückseite gezeigt wird, während der Deckel auf beiden Aufnahmen von der gleichen Seite zu sehen ist.

93. Als schwächere Variante von Nr. 3 könnte möglicherweise eine Flöte im Museum Boymans–van Beuningen in Rotterdam angesprochen werden (Inv. Nr. 7). Nr. 5 wandelt das durch Oskar Rauter überlieferte Glas I 3g (Abb. 6) ab. Nr. 6 ist dem New Yorker Glas I 8 (Abb. 11) verwandt. Nr. 15 variiert das Schema des Pokals I 6a (Abb. 9).

gefertigten Gläsern beträchtliche Qualitätsschwankungen zu verzeichnen sind. Allerdings dürfte bei den Unregelmässigkeiten der Ausführung auch das Bemühen eine Rolle spielen, durch den Verzicht auf technische Perfektion dem lebendigen Eindruck der Gläser des 17. Jahrhunderts nahezukommen.[94] Gelegentlich ist ein Mangel an Verständnis für die Eigenart venezianischer Glaskunst nicht zu übersehen. So offenbart die Ausstaffierung des Glases Nr. 20 (Abb. 53) mit völlig unorganisch angefügten symmetrischen Kämmen und gekniffenen Ansätzen Müllers gestalterische Unsicherheit.[95] Das aus diesem Typ entwickelte Glas der Gruppe III (Abb. 49) dagegen belegt ein weiteres Mal die Qualität und Eigenständigkeit, die Müllers Gläser durch die Zusammenarbeit mit Kunstgewerbeentwerfern wie Bichweiler und Keller-Leuzinger gewonnen haben.

Bei einigen der auf den Musterfotos gezeigten Stücke sind bereits Ansätze zur Loslösung von den überlieferten Typen des 17. Jahrhunderts festzustellen. Gläser wie die Flöte Nr. 12 (Abb. 54) verdanken ihre Entstehung allerdings eher der Tradition der Kleinbläserei des Thüringer Waldes als dem Geist des neuen kunstgewerblichen Schaffens. Bemerkenswert sind dagegen das Reichsadlerglas Nr. 19 (Abb. 53) und der Kettenschlangenpokal Nr. 24 (Abb. 54), die—hervorgehoben auch durch die Höhe des Preises—als Bravourstücke der Lampentechnik gewertet werden müssen und sicher als solche gedacht waren.

---

94. Auffallend ist, dass die verschmolzenen Brüche, die verschiedentlich an den Schäften der erhaltenen Gläser zu beobachten sind, bereits auf den Musterfotos festgestellt werden können; es handelt sich also um Korrekturen während des Herstellungsprozesses.

95. Vgl. auch den Blütenpokal I 12b (Abb. 20).

English Summary

The article is meant as a contribution to our knowledge of nineteenth-century glass as well as to the problems arising in connection with certain glasses made in the Venetian style. The dragon-stemmed goblets discussed here have been classified nearly without exception as works of sixteenth or seventeenth centuries. In order to identify the glasses in question as products of the second half of the nineteenth century, the usual approach of a stylistic analysis is not made, but rather a consideration of manufacturing techniques leads to the conclusion that the glasses were blown at the lamp.

It is asserted that the maker of these glasses was Carl Heinrich Florenz Müller (1845–1912), who, originating from Pisau in the Thuringian Forest, founded a workshop for artistic glassblowing in Hamburg in 1865. Müller produced art glasses in the Venetian style from 1865 up to the beginning of the 1880's. Later he concentrated on manufacturing electrical equipment and produced the first x-ray tubes in Germany.

According to notes of Justus Brinckmann, the first director of the Hamburg Museum für Kunst und Gewerbe, Müller worked in his early years for dealers from the Netherlands who used the results of his skill for fraudulent purposes. In fact nearly all the Müller glasses in public collections were declared to be Venetian goblets from the sixteenth century when they were purchased. The greater part of Müller's glasses belong to the category of copies and imitations, referred to in the article as Group I. As these glasses are quite heavy (and occasionally up to half a meter high) the possibility that they might not be blown at the furnace has never been discussed. But there are several features which indicate that their origin was at the lamp:

1. The glasses do not show a pontil mark. Objects at the lamp are handled by use of special clamps and pincers.

2. The rim of the foot is folded upward. It moves this direction automatically in course of working.

3. The surface of the bowls is slightly irregular. A large form can be blown at the lamp only step by step.

4. The rods used for the stem are very regularly built. Prefabricated material, produced by special firms, was used.

5. There is an opening on the inside of the covers, at the top, caused by their being blown from a tube.

6. Some decorative motifs, such as the combs on the dragon heads, are produced from separate trailings of glass threads. At the furnace, similar details are produced by cutting a pad of glass with a knife or shears.

There are several reasons why lamp-blown glasses like Müller's cannot originate from the sixteenth or seventeenth centuries. From medieval, perhaps even from Roman times, glass was blown using oil-fed lamps. These lamps allowed only rather localized heating. The numerous technical treatises on lamp-blowing, the first being Johan Kunckel's *Ars Vitraria*, show that lamp-blowing in the seventeenth and eighteenth centuries was restricted to producing minor objects like pearls, small statuettes, toys, dollhouse fittings, etc., and for scientific equipment like the instruments made about 1660 for the Academia del Cimento in Florence.

Lamp-blown glasses of greater size and weight only can be blown by use of gas burners which supply the blower with a more steady and intense heat. These burners were not developed before the middle of the nineteenth century, so the invention of this new technique can be taken as a *terminus post quem*.

About 1876/77 Müller stopped producing imitations of Venetian glasses that could be sold as authentic. He contacted two renowned applied art specialists, Franz Keller-Leuzinger and Robert Bichweiler, whose designs he carried out. The result of this collaboration are the glasses of Group II, consisting of vessels to be used for drinking, freely designed in Venetian style. These glasses are of high quality and cannot be identified as anything other than products of the second half of the nineteenth century.

The goblets of Group III show the technical and stylistic innovations of Group II adapted to the style of the glasses of Group I. For example, the ribbed feet and baluster stems resemble those in the glasses in Group II. From the point of view of technical quality and artistic value these glasses must be considered the most important group of historistic glass made in Germany. They can be compared only with the art glass of Köln-Ehrenfeld, starting in the early 1880's when Müller's production ended.

KATALOG

Stücke, deren Ausführung durch C. H. F. Müller dokumentarisch nachzuweisen oder deren Entstehung im 19. Jahrhundert belegt ist, sind mit einem Stern (*) gekennzeichnet.

Bei den Literaturangaben wurde jeweils nur die neueste oder die am besten illustrierte Publikation angeführt.

Gruppe I
Nachbildungen und Stilimitationen

I 1a

Abb. 2

Deckelpokal

C. H. F. Müller, Hamburg, um 1865/70; H. 40,3 cm

Konische Kuppa mit gerundetem Ansatz. Dünner massiver Stengelschaft, am Kuppaansatz leicht verdickt. Flacher Fuss mit nach oben umgeschlagenem Rand, keine Abschlagnarbe. Sehr klares farbloses Glas ohne Lufteinschlüsse. Schafteinsatz mit geschwungener Kontur und Doppelacht aus kaum merklich tordierten Stäben mit zwei gegenläufig spiralig geführten Einlagen aus voneinander getrennten blauen, weissen, roten und gelben Fäden. Kämme und Ansätze blaugrün, die schnabelartigen Fortsätze unterhalb der Kämme kräftig, doch wenig bewegt gestaltet. Deckel annähernd halbkugelförmig, die Bekrönung analog dem Schafteinsatz gebildet, zwischen den ungeschwungenen Kämmen Türmchen mit Querrippung. Verwendung von Zangen mit kleinteiligem Waben- und Rautenmuster, die Prägung durch zu starkes Erhitzen leicht zerschmolzen. Am Fussansatz Spannungsrisse.

London, Victoria & Albert Museum. Inv. Nr. 579–1903

Lit.: W. B. Honey, *Glass, A Handbook . . .*, London 1946, pl. 28A

I 1b

Pokal (Deckel nicht zugehörig, vgl. I 6b)

C. H. F. Müller, Hamburg, um 1865/70; H. 28,5 cm

Ähnlich I 1a. Abweichend: kleinere, geringer geschwungene Kämme, die oberen Schnabelteile länger und stärker bewegt. Das Waben- und Rautenmuster der Zangen weniger stark zerschmolzen.

Cambridge, Fitzwilliam Museum. Inv. Nr. Mar C 138–1912

Marley Bequest 1912

I 1c

Deckelpokal

C. H. F. Müller, Hamburg, um 1865/70; H. 42,5 cm

Ähnlich I 1a. Abweichend: gedrückte Deckelform, schmalere Kämme als Bekrönung des Schafteinsatzes. Fadeneinlagen weiss-gelb. Ausschliessliche Verwendung einer Zange mit Parallelriefenmuster.

State College, Pennsylvania. Jerome Strauss Collection. Nr. S 1371

Lit.: Ausst. Kat. *Glass Drinking Vessels from the Strauss Collection*, The Corning Museum of Glass, Corning, N.Y., 1955, Nr. 120, S. 49/50 mit Abb.

I 1d

Pokal

C. H. F. Müller, Hamburg, um 1865/70; H. 26,8 cm

Sehr ähnlich I 1c. Schnabelfortsätze differierend, vgl. I 1b. Deckel verloren.

State College, Pennsylvania. Jerome Strauss Collection, Nr. S 1937

I 1e

Deckelpokal

C. H. F. Müller, Hamburg, um 1865/70; H. 44 cm

Ähnlich I 1a. Fadeneinlagen rot, gelb, weiss; Ansätze blau. — Bekrönung zwischen den Deckelflügeln (Türmchen?) verloren.

Paris, Palais des Beaux Arts de la Ville (Petit Palais). Inv. Nr. 13092

Legat Dutuit 1902

I 1f

Deckelpokal

C. H. F. Müller, Hamburg, 70er Jahre 19. Jh.; H. 39,4 cm

Ähnlich I 1a. Abweichend: Geringere Unregelmässigkeiten in der Ausführung. Ungewöhnlich hoher, halbkugelförmiger Deckel. Schafteinsatz und Deckelbekrönung aus dünneren Stäben. Zwischen den

Deckelflügeln ein Vogel anstelle des Türmchens. Schnäbel flach und nach auswärts geschwungen. Verwendung zweier Zangentypen mit feiner und gröberer Waffelmusterung (Quadrate).

Arnhem, Gemeentemuseum. Inv. Nr. VS 159
Leihgabe aus Privatbesitz

***I 1g**
Abb. 1

Deckelpokal

C. H. F. Müller, Hamburg, 70er Jahre 19.Jh.; H. 49 cm

Ähnlich I 1f. Abweichend: gedrückte Deckelform. Stäbe des Schafteinsatzes und der Deckelbekrönung leicht tordiert. Fadeneinlagen in Weiss und Gelb. Die dünn ausgezogenen Schnäbel s-förmig geschwungen. Stengelansatz der Kuppa stark, der des Fusstellers leicht verdickt. Zangenmusterung wie I 1d. — Mehrfach geklebt, Teile der Ansätze und der Schaftverzierung verloren. Für den ursprünglichen Zustand vgl. Abb. 53, Nr. 21.

Hamburg, Museum für Kunst und Gewerbe. Inv. Nr. 1893–71. 1893 von C. H. F. Müller für 40,– Mark erworben. Original verschollen.
Lit.: *Röntgenstrahlen* (Anm.8), S.43, Farbabbildung

I 1h

Pokal

C. H. F. Müller, Hamburg, 70er Jahre 19.Jh.; H. 25,4 cm

Ähnlich I 1g. Abweichend: Schafteinsatz in der Proportionierung etwas schlanker, Stäbe stark tordiert, Fuss- und Kuppaansatz kräfiger betont.

Den Haag, Gemeentemuseum. Inv. Nr. OG-96-54
Legat W. J. H. Mulier 1954
Lit.: *Catalgus van Noord- en Zuidnederlands Glas*, Gemeentemuseum Den Haag, 1962, Nr. 18, S.25 mit Abb.

I 1i

Pokal

C. H. F. Müller, Hamburg, 70er Jahre 19.Jh.; H. 30 cm

Sehr ähnlich I 1g. Abweichend: In der Glasmasse feine Bläschen. Fadeneinlagen weiss-rot-blau-gelb, Schnabelendungen stark abgeknickt.

Frankfurt, Museum für Kunsthandwerk.
Inv. Nr. 7007
Schenkung Kessler-Kollings 1934
Lit.: Katalog *Europäisches und aussereuropäisches Glas*.

ABB. 1. *C.H.F. Müller, Hamburg, 70er Jahre 19.Jh.: Hamburg, Museum für Kunst und Gewerbe. Ankauf vom Künstler. Original verschollen (I 1g)*

ABB. 2. *C.H.F. Müller, um 1865/70: London, Victoria & Albert Museum (I 1a)*

Bearbeitet von A. Ohm. Museum für Kunsthandwerk, Frankfurt/M. 1973, Nr. 188, S.88 mit Abb.

I 1k

Pokal

C. H. F. Müller, Hamburg, 70er Jahre 19.Jh.; H. 24,5 cm

Ähnlich I 1g. Kämme etwas stärker einwärts geschwungen, kräftigere Verdickung am Fussansatz. Fussteller gesprungen.

Auktion Galerie Fischer Luzern, 7.Nov.1949, *Glassammlung N.*, Nr. 512 mit Abb. Verbleib unbekannt.

I 2a

Einfaches Acht-Glas mit Deckel

C. H. F. Müller, Hamburg, um 1865/75; H. 35,2 cm

Stengeleinsatz und Deckelaufsatz, welche die Form einer 8 zeigen, sind mit gelben eingeschmolzenen Spiralfäden und blaugrünen seitlichen Ansätzen versehen (Inventareintrag).

Hamburg, Museum für Kunst und Gewerbe.
Inv. Nr. 1893–70
1893 von C. H. F. Müller für 26 Mark erworben. Laut Inventarvermerk *total zerbrochen*. Verschollen.

I 2b
Abb. 3

Pokal

C. H. F. Müller, Hamburg, 1865/70; H. 21,2 cm

Sehr klares farbloses Glas ohne Lufteinschlüsse. Schaftverzierung: gedrückte Tropfenform mit eingefügter 8, unverbundene weiss-rot-weisse Fadeneinlagen. Kämme und Ansätze blaugrün. Anstelle der Schnabelfortsätze eine brückenartige Verbindung zwischen den Kämmen und den Stäben des Schafteinsatzes. Waben- und Rautenmuster der Zangen wie 1a, b, desgl. die verschmolzene Oberfläche.

Frankfurt, Museum für Kunsthandwerk.
Inv. Nr. 7008 b
Schenkung Kessler-Kollings 1934

Lit.: Kat. Frankfurt (vgl. I 1c), Nr. 190, S.89 mit Abb.

I 2c
Abb. 3

Pokal

C. H. F. Müller, Hamburg, 70er Jahre 19. Jh.; H. 20,6 cm

Ähnlich I 2b. Abweichend: Schaftverzierung schlanke Tropfenform, Fadeneinlage in gelb-weiss-rot-blau, bandartig aneinandergefügt. Verwendung von Zangen mit feinem Waffelmuster (Quadrate).

Frankfurt, Museum für Kunsthandwerk.
Inv. Nr. 7008 a
Schenkung Kessler-Kollings 1934

Lit.: Kat. Frankfurt (vgl. I 1c), Nr. 190, S.89 mit Abb.

I 2d

Pokal

C. H. F. Müller, Hamburg, 70er Jahre 19. Jh.; H. 24 cm

Ähnlich I 2b. Abweichend: die helle Glasmasse von kleinen Bläschen durchsetzt. Schnäbel unterhalb der Kämme entsprechend I 1c. Stäbe des Schafteinsatzes stark tordiert. Fadeneinlagen wie I 2c. Schaft am Fussansatz verdickt. — Schaft zwischen Kuppa und Einsatz gebrochen. Kuppa geklebt.

Lüttich, Musée du Verre. Inv. Nr. B.563
Slg. Baar; 1885 in Lüttich erworben

I 2e

C. H. F. Müller, Hamburg, 1865/70; H. ca. 27 cm

Ähnlich I 2b. Kuppa in Form eines Spitzkelchs. Einer der Kämme verloren. Worms, Stiftung Kunsthaus Heylshof.

I 3a
Abb. 5

Pokal

C. H. F. Müller, Hamburg, um 1865/70; H. 24,5 cm

Glasmasse, Kuppa- und Fussform wie I 2b. Als Schaftverzierung einfache Tropfenform ohne Innenfigur. Ansätze und Fadendekor wie I 1a. Verwendung von Zangen mit feinem Waffelmuster, Schnabelfortsätze leicht geschwungen. — Kuppa und Schafteinsatz durch Metallklammer verbunden.

Amsterdam, Rijksmuseum. Inv. Nr. NM 19754–10
Legat. A. J. Enschedé 1896

I 3b

Pokal

C. H. F. Müller, Hamburg, um 1865/70; H. 25,4 cm

Ähnlich I 3a. Etwas weniger sorgfältig gearbeitet. Abweichend: Kuppa steilwandiger und geringfügig kleiner, Kämme etwas schmaler. Stab des Schafteinsatzes wenig tordiert, so dass Farbfäden sich nicht zu eng geführter Spirale ordnen.

Cambridge, Fitzwilliam Museum.
Inv. Nr. C 140–1912
Marley Bequest 1912

I 3c
Abb. 4

Pokal

C. H. F. Müller, Hamburg, 70er Jahre 19. Jh.; H. 25 cm

Ähnlich I 3a. Abweichend: In der Glasmasse feine Bläschen. Höhere Kuppa. Anstelle der Kämme Bekrönung aus übereinandergeschichteten blaugrünen Fäden. Material des Einsatzes leicht gelbstichig, stark

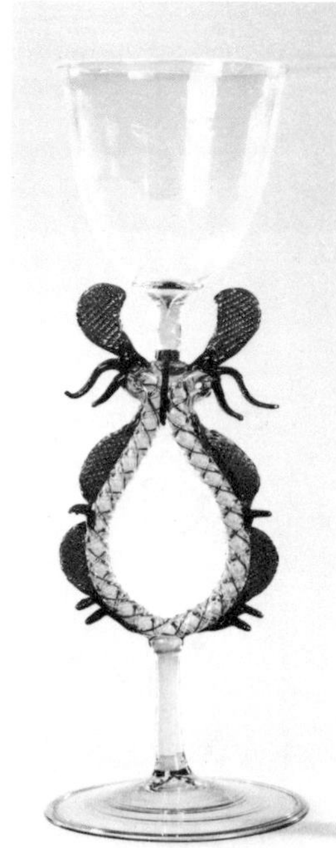

tordiert. Fadeneinlage bandartig blau-rot-weiss-gelb wie I 2c und andere. Schaft am Fussansatz verdickt. — Der untere Teil des Schaftes gebrochen, die Bruchstelle verschmolzen.

Frankfurt, Museum für Kunsthandwerk.
Inv. Nr. 7005 a
Schenkung Kessler-Kollings 1934
Lit.: Kat. Frankfurt (vgl. I 1i), Nr. 191, S. 89 mit Abb.

I 3d
Abb. 4

Pokal

C. H. F. Müller, Hamburg, 70er Jahre 19. Jh.;
H. 24,7 cm

Sehr ähnlich I 3c. Abweichend: Keine Schaftverdickung am Fussansatz. Schafteinsatz kaum tordiert. Farbfäden nicht bandartig verschmolzen. Kämme weniger sorgfältig geschichtet.

Frankfurt, Museum für Kunsthandwerk.
Inv. Nr. 7005 b
Schenkung Kessler-Kollings 1934
Lit.: Kat. Frankfurt (vgl. I 1i), Nr. 191, S. 89 mit Abb.

I 3e

Pokal

C. H. F. Müller, Hamburg, 70er Jahre 19. Jh.;
H. 22,5 cm

Abb. 3. *C. H. F. Müller, links 70er Jahre 19. Jh., rechts um 1865/70; Frankfurt, Museum für Kunsthandwerk, Schenkung Kessler-Kollings (I 2c, b)*

Abb. 4. *C. H. F. Müller, 70er Jahre 19. Jh.; Frankfurt, Museum für Kunsthandwerk, Schenkung Kessler-Kollings (I 3c, d)*

Abb. 5. *C. H. F. Müller, um 1865/70; Amsterdam, Rijksmuseum, Legat Enschedé (I 3a)*

Ähnlich I 3c. Abweichend: Kuppa etwas konischer gebildet. Kämme aus weniger Fäden locker geschichtet. Fäden des Schafteinsatzes rot und weiss.

Auktion 555, Kunsthaus Lempertz, Köln, Herbst 1976, Kat. Nr. 1315 mit Abb. Verbleib unbekannt.

I 3f

Pokal

In der Art des C. H. F. Müller, 70er Jahre 19. Jh.;
H. 26,7 cm

Ähnlich I 3c. Die Tropfenform des Einsatzes etwas gestreckter. Insgesamt nicht ganz so sauber gearbeitet. Fadeneinlagen bandartig blau-rot-weiss. Abrissimitation durch angeschmolzenes Bruchstück eines Glasstabs. — Die Kämme und die oberen Schnabel-

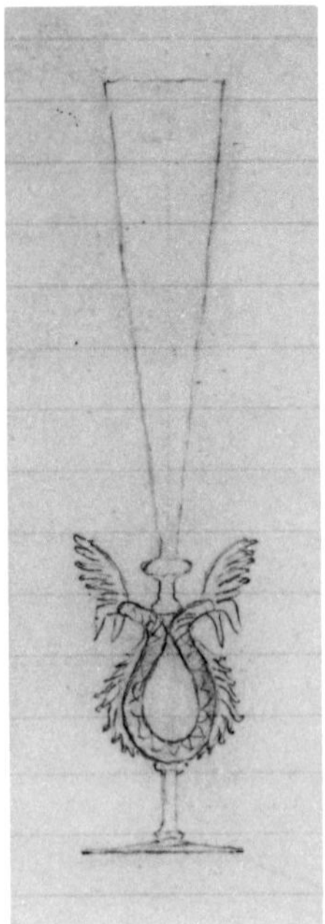

Abb. 6. *Zeichnung nach einer vermutlich von C.H.F. Müller geblasenen Flöte. Aus einem Skizzenbuch Oskar Rauters von 1881 im Kunstmuseum Düsseldorf; das Original verschollen (I 3g)*

Abb. 7. *C.H.F. Müller, 70er Jahre 19.Jh.; Frankfurt, Museum für Kunsthandwerk, Schenkung Kessler-Kollings (I 4a, b)*

fortsätze sowie ein Teil der gekniffenen Ansätze fehlen.

Dresden, Kunstgewerbemuseum Schloss Pillnitz.
Inv. Nr. 43411

I 3g
Abb. 6

Flöte

C. H. F. Müller?, Hamburg?, 70er Jahre 19.Jh.;
H. 29 cm

Variante des Typs I 3. Anstelle der Pokalkuppa spitze Flöte mit Nodus. Die Aussenseiten der Schaftverzierung mit schuppenartig übereinandergeschmolzenen Glasstäbchen besetzt. Die Kämme wohl geschichtet ähnlich I 3c.

Überliefert als Zeichnung in einem Skizzenbuch Oskar Rauters von 1881 in Kunstmuseum Düsseldorf. Verbleib unbekannt.

I 4a
Abb. 7

Pokal

C. H. F. Müller, Hamburg, 70er Jahre 19.Jh.;
H. 30,7 cm

Glasmasse, Kuppa- und Fussform ähnlich I 3c. Schafteinsatz in unregelmässiger Tropfenform konturiert. Innenform vierfach geschlungen. Voneinander getrennte weisse, rote, gelbe und blaue Fäden. Wenig sorgfältig gearbeitet. — Die wohl nicht zugehörige Kuppa (überlanger Schaft) nachträglich angeschmolzen. Die Endung eines der unteren Schnabelfortsätze und der Auslauf eines der seitlichen Ansätze verloren.

Frankfurt, Museum für Kunsthandwerk.
Inv. Nr. 7012
Schenkung Kessler-Kollings 1934
Lit.: Kat. Frankfurt (vgl. I 1i), Nr. 189, S.89 mit Abb.

I 4b
Abb. 7

Pokal

C. H. F. Müller, Hamburg, 70er Jahre 19.Jh.;
H. 22,8 cm

Besser gearbeitete Variante von I 4a. Abweichend: Kämme mit Waffelmuster anstelle der geschichteten Stäbchen. Anordnung des oberen Teils der Innenfigur abgewandelt, Farbfäden der Schaftverzierung bandartig verschmolzen. Stengel am Ansatz des leicht gewölbten Fusstellers nicht verdickt. — Verschmolzener Bruch zwischen Kuppa und Schafteinsatz.

Frankfurt, Museum für Kunsthandwerk.
Inv. Nr. 7011
Schenkung Kessler-Kollings 1934
Lit.: Kat. Frankfurt (vgl. I 1i), Nr. 187, S.88 mit Abb.

I 5a

Pokal

C. H. F. Müller, Hamburg, 70er Jahre 19.Jh.;
H. 24,8 cm

Glasmasse, Kuppa- und Fussform ähnlich I 3c. Schafteinsatz annähernd brezelförmig angelegt. Voneinander getrennte weiss-rot-gelb-blaue Fäden. Die blaugrünen Ansätze mit feinem und gröberem Waffelmuster versehen. Die Schnabelfortsätze ähnlich I 1i. — Ein ursprünglich vorhandenes Pendent wurde abgegeben.

Frankfurt, Museum für Kunsthandwerk.
Inv. Nr. 7009

Schenkung Kessler-Kollings 1934
Lit.: Kat. Frankfurt (vgl. I 1i), Nr. 186, S.88 mit Abb., s. auch G. Weiss, Ullstein Gläserbuch, Berlin/Frankfurt/Wien 1966, Abb. S.118.

I 5b
Abb. 8

Deckelpokal

C. H. F. Müller, Hamburg, 70er Jahre 19. Jh.; H. 45 cm

Ähnlich I 5a. Die Innenschlinge und die Schnabelfortsätze des Schafteinsatzes geringfügig abweichend. Kräftig gewölbter Deckel mit Wiederholung des Schaftmotivs, bekrönt von einem Vogel. Fadeneinlagen gelb und rot, Ansätze dunkelblau. — Teile der Schnäbel am Deckel und der Ansätze der Schaftverzierung verloren.

Lund, Kulturen (Kulturhistoriska Museet).
Inv. Nr. KM 3.993.
1886 in Kopenhagen erworben

Lit.: *Kulturen. En Årsbok 1953.* Glas. Lund 1954, S.59 mit Abb.

I 5c

Pokal

C. H. F. Müller, Hamburg, vor 1876; H. 17,1 cm

Im Aufbau ähnlich I 5a, b, doch gröber gearbeitet (vgl. I 10g, h und Abb. 17). Im ganzen kleiner. Verwendung stärkerer Stäbe. Fadeneinlagen gelbweiss. — Schaft unterhalb des Ziereinsatzes gebrochen.

Sehr ähnlich Abb. 53, Nr. 7
Dresden, Kunstgewerbemuseum Schloss Pillnitz.
Inv. Nr. 3354
Erworben 1876 von dem Amsterdamer Händler Goudstikker

I 6a
Abb. 9

Deckelpokal

C. H. F. Müller, Hamburg, um 1865/75; H. 46,4 cm

Glasmasse, Kuppa- und Fussform ähnlich I 2b. Schafteinsatz in Form einer vielfach gewundenen Schlange mit kleinen gekniffenen Ansätzen an den äusseren Schlingungen und gezacktem Kamm an Hals und Kopf. Einlage aus unverbundenen Fäden in weiss-rot-gelb-blau. Ansätze blaugrün. Verwendung einer Zange mit feinem Waffelmuster. Deckel in Form einer gedrückten Halbkugel, als Bekrönung vereinfachte Wiederholung des Schafteinsatzes.

Abb. 8. *C. H. F. Müller, 70er Jahre 19. Jh.; Lund, Kulturen (I 5b)*

Abb. 9. *C. H. F. Müller, um 1865/70; Amsterdam, Rijksmuseum, Legat Enschedé (I 6a)*

Amsterdam, Rijksmuseum. Inv. Nr. NM 10754:2
Legat A. J. Enschedé 1896

Lit.: H. E. van Gelder, *Glas en Ceramiek. De Kunsten van het Vuur.* Utrecht 1955, Tf. IX, 3

I 6b

Deckel

C. H. F. Müller, Hamburg, um 1865/75; H. 15,5 cm

Glasmasse und Farbe der Fäden und Ansätze wie I 6a. Die Deckelform etwas gedrückter. Die geknäulte Schlangenfigur gestreckter und ohne seitliche ge-

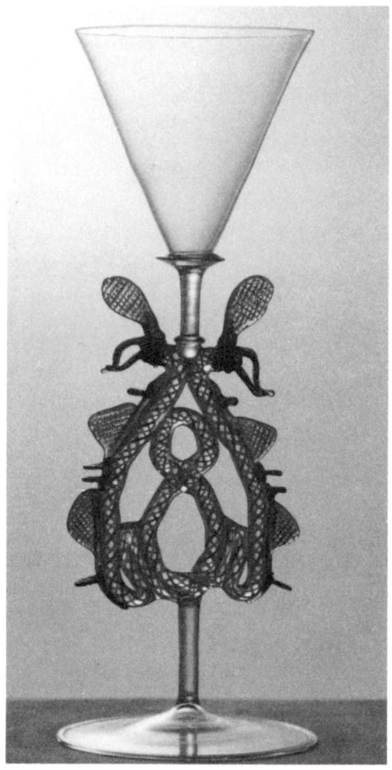

ABB. 10. *C.H.F. Müller, um 1865/70; London, British Museum, Slade Bequest (I 7a)*

ABB. 11. *C.H.F. Müller, um 1865/70; New York, Metropolitan Museum (I 8)*

kniffene Ansätze. — Oberer Teil des schnabelartigen Rachens verloren.

Cambridge, Fitzwilliam Museum. Inv. Nr. Mar C 138–1912

Marley Bequest 1912

I 6c

Deckel und Schaftfragment eines Pokals

C. H. F. Müller, Hamburg, um 1865/75; H. 16,4 cm (Deckel); DM. 8,8 cm (Schaftfragment)

Von einem Pokal ähnlich I 6a. Die Ansätze gekniffen mit Parallelriefenmusterung. — Einige Spitzen am Kamm der Schlange der Deckelbekrönung verloren. Vom Pokal nur der stark gewundene Schafteinsatz ohne Schlangenhals und -kopf erhalten.

Hannover, Kunsthandel Uwe Friedleben

I 7a
Abb. 10

Deckelpokal

C. H. F. Müller, Hamburg, um 1865/70; H. 49,5 cm

Kuppa- und Fussform ähnlich I 2b. Kuppa etwas höher. Stab des Einsatzes leicht tordiert. Farbe der unverbundenen Fäden rot, weiss und gelb. Geschwungene Kämme mit nur leicht bewegten Schnabelfortsätzen ähnlich I 1a. Verwendung einer feinen und einer gröberen Zange mit Waffelmuster. Deckelform ähnlich I 1a. Der Aufsatz verfeinerte Wiederholung der Schaftverzierung. Als Bekrönung ein Vogel zwischen ungeschwungenen Kämmen.

London, British Museum, Slade Bequest, Nr. 589, 1868

Lit.: A. Nesbitt, *Catalogue of the Collection of Glass Formed by Felix Slade, Esq.* (London 1871)

I 7b

Pokal

C. H. F. Müller, Hamburg, um 1865/70; H. 29 cm

Ähnlich I 7a. Fadeneinlagen gelb, weiss, rot; Ansätze blau.

Paris, Palais des Beaux-Arts de la Ville (Petit Palais). Inv. Nr. 1309
Legat Dutuit 1902

I 7c

Pokal

C. H. F. Müller, Hamburg, um 1865/70; H. 28,3 cm

Schaft und Fuss ähnlich I 7a, doch weniger sorgfältig gearbeitet. Kuppa in Form eines Spitzkelchs, durch eine Scheibe vom Schaft abgesetzt. Die Kämme ohne Schwingung, entsprechend der Deckelbekrönung von I 7a. Farbe der Fäden gelb, rot, blau, weiss. Verwendung einer Zange mit feinem Waffelmuster.

Cambridge, Fitzwilliam Museum.
Inv. Nr. Mar. C 137–1912
Marley Bequest 1912

I 7d

Pokal

C. H. F. Müller, Hamburg, um 1865/75; H. 29,8 cm

Ähnlich I 7a, in der Gesamtwirkung gestraffter. Der leicht gelbstichige Stab des Einsatzes tordiert, die zum Teil miteinander verschmolzenen Fäden gelb, rot, weiss, blau. Die mit einer feinen und einer gröberen Waffelmusterzange geformten Ansätze blaugrün.

Nürnberg, Germanisches Nationalmuseum.
Inv. Nr. G1 93
Slg.Felix 1886

Lit.: Kat. Auktion *Slg.Felix*, Kunsthaus Lempertz. Köln, Okt. 1886, Nr. 169, S. 33 (beschrieben als Deckelpokal des 16./17. Jh., H. 45,5 cm).

I 7e

Fragmente eines Deckelpokals

C. H. F. Müller, Hamburg, vor 1879;
H. ehm.ca. 41 cm

Ähnlich I 7d, der Stab der Schaftverzierung schwächer tordiert. Fäden gelb-weiss, bandartig verschmolzen. Als Innenfigur der Bekrönung anstelle der drei Kreisformen ein spiralförmig gedrehter Stab (vgl. Abb. 53, Nr. 18). — Ein grosser Teil des Fusstellers, der obere Teil des Schaftes mit dem Kuppaanschluss sowie alle Kämme und die Flügel des Vogels der Deckelbekrönung verloren.

Dresden, Kunstgewerbemuseum Schloss Pillnitz.
Inv. Nr. 8201

Erworben von Antiquar Hugo Klein, Dresden, Jan. 1879

I 8
Abb. 11

Pokal

C. H. F. Müller, Hamburg, um 1865/70; H. ca. 28 cm

Die Form des Spitzkelchs der Kuppa und des Fusses ähnlich I 7c, die Tropfenform des Schafteinsatzes schlanker. Als Innenfigur eine 8 flankiert von zwei zusammengelegten Schlaufen. Dreiecksförmige Ansätze und Kämme ohne Schwingung.

New York, The Metropolitan Museum of Art.
Inv. Nr. 27.185.301
1927 Museum Fund

Lit.: The Metropolitan Museum of Art. *A Special Exhibition of Glass from the Museums Collections*, New York 1936, Fig. 23

I 9a

Deckelpokal

C. H. F. Müller, Hamburg, 1865/70; H. 45,7 cm

Form von Kuppa, Fuss und Deckel ähnlich I 1a, die Ausführung jedoch sorgfältiger, insgesamt besser proportioniert. Die Schaftverzierung gebildet aus räumlich geführten, spiralig gegeneinandergedrehten Doppelstäben. Gekniffene Ansätze und geschwungene Kämme mit kurzen, wenig bewegten Schnabelfortsätzen. Die Deckelbekrönung mit Vogel und flankierenden ungeschwungenen Kämmen analog I 1f, Zangen mit Waben- und feinem Waffelmuster. Die Oberfläche der gekniffenen Partien leicht überschmolzen. — Geringfügige Ausbrüche bei den Ansätzen der Schaftverzierung.

Hartford, Conn., The Wadsworth Atheneum.
Inv. Nr. 1917, 334
Ehm. Slg. J. Pierpont Morgan 1917

Lit.: A. von Saldern, Spiral-Schlangenpokale à la façon de Venise in: *Jahrbuch der Hamburger Kunstsammlungen*, Bd. 17, 1972, S.73 f., Abb. 10 (Nr. 13)

I 9b
Abb. 12

Deckelpokal

C. H. F. Müller, Hamburg, 1865/70; H. 52,5 cm

Formal sehr ähnlich I 9a. Abweichend: Fadeneinlagen weiss. Blaue Ansätze, Verwendung von Zangen mit feinem und gröberem Waffelmuster. Ehm. Wien Österreichisches Museum für angewandte Kunst. Inv. Nr. 21312. Verschollen.

Lit.: Br. Bucher, *Die Glassammlung des K. K. Oesterreich.* Museum, Wien 1888, S.83. Abb.: *Das Kunsthandwerk. Sammlung mustergültiger kunstgewerblicher Gegenstände aller Zeiten*, hrsg. von Br. Bucher A. Gnauth, Jg. II, Stuttgart 1875, Tf. 46a

I 9c
Abb. 13

Deckelpokal

C. H. F. Müller, Hamburg, 1865/70; H. 49 cm

Formal sehr ähnlich I 9a, b. Abweichend: nur geringfügig gegeneinander verdrehte Einlagen in einzeln geführten weiss, gelb, rot und blau gefärbten Fäden. — Einer der Kämme der Schaftverzierung abgebrochen, aber noch vorhanden. Teile der Deckelansätze verloren. Kuppa gebrochen und geklebt.

Amsterdam, Rijksmuseum.
Inv. Nr. NM 10754:1 a/b
Legat A. J. Enschedé 1896

ABB. 12. *C. H. F. Müller, um 1865/70; aus* Das Kunsthandwerk, *Bd. II, Stuttgart 1875. Original im Österreichischen Museum für angewandte Kunst in Wien; heute verschollen (I 9b)*

ABB. 13. *C. H. F. Müller, um 1865/70; Amsterdam, Rijksmuseum, Legat Enschedé (I 9c)*

I 9d

Pokal

C. H. F. Müller, Hamburg, 1865/70; H. 26 cm

Variante von I 9 a-c. Abweichend: die Doppelstäbe des Schafteinsatzes zu einem dicken, tauartigen Strang umeinandergewunden. — Kämme und drei oder vier Schnabelfortsätze verloren, Kuppa und Schaftverzierung durch Metallmanschette verbunden.

Lüttich, Musée du Verre. Inv. Nr. B. 1218
Slg. Baar; 1909 in Amsterdam erworben
Lit.: v. Saldern (vgl. I 9a), S. 73, Nr. 11 ohne Abb.

I 9e

Pokal

C. H. F. Müller, Hamburg, vor 1874; H. 32,8 cm (mit dem verlorenen Deckel ehm. 57 cm)

Ähnlich I 9 a-c. Abweichend: Verdickung des Schaftes am Fussansatz. Dünne, an den Enden scharfwinklig abgebogene Schnabelfortsätze (vgl. I 3c, d). Bandartig verschmolzene Fadeneinlagen weiss-gelb-blau-rot. — Zwei Schnabelendungen verloren. Kleiner Ausbruch an einem Ansatzstück. Schaft am Ansatz der Zierfigur gebrochen und geklebt, darüber ältere, verschmolzene Bruchstelle.

Nürnberg, Gewerbemuseum der Landesgewerbeanstalt Bayern. Inv. Nr. 2961/1
Erworben 1874 aus der Slg. Völkner, Prag

I 10a
Abb. 14

Deckelpokal

C. H. F. Müller, Hamburg, 70er Jahre 19. Jh.; H. 43 cm

Im Aufbau ähnlich den Gläsern des Typs I 9. Anstelle der geschwungenen Kämme hohlgeblasener Drachenkopf und spiralig gewundener, farbloser, von blauem Faden konturierter Schwanz. Form von Kuppa und Deckel ähnlich I 9 a-c. Schaft am Fussansatz verdickt. Ansätze blaugrün, Fäden gelb und rot. — Teile der Deckelbekrönung, der Schaftverzierung und des Drachenschwanzes verloren.

Paris, Musée Jacquemart-André, Inv. Nr. D 544 I 754
Lit.: v. Saldern (vgl. I 9a) S. 73, Abb. 9 (Nr. 12)

I 10b
Abb. 15

Deckelpokal

C. H. F. Müller, Hamburg, 70er Jahre 19. Jh.; H. 48,7 cm; ohne Deckel 31,4 cm

Farbloses Glas mit vereinzelten Bläschen, leichter Graustich. Fadeneinlagen gelb-weiss-gelb, begleitet von weissem Einzelfaden. Die Ansätze blau, mit feinem Waffelmuster. Leichte Verdickung des Schaftes am Ansatz des Fusses. — Die Deckelbekrönung und der Kopf des Drachens der Schaftverzierung Kriegsverlust. Schaft unterhalb der Kuppa gebrochen und verschmolzen.

ABB. 14. *C.H.F. Müller, 70er Jahre 19.Jh.; Paris, Musée Jacquemart-André (I 10a)*

ABB. 15. *C.H.F. Müller, 70er Jahre 19.Jh.; Hamburg, Museum für Kunst und Gewerbe, Zustand 1930, vor Kriegsbeschädigung (I 10b)*

ABB. 16. *C.H.F. Müller oder Nachahmer, 70er Jahre 19.Jh.; London, Victoria & Albert Museum, Slg. Buckley (I 10f)*

Hamburg, Museum für Kunst und Gewerbe.
Inv. Nr. 1930–71
1930 aus dem Hamburger Kunsthandel erworben
Lit.: v. Saldern (vgl. I 9a), S.71 f., Abb. 6 (Nr. 8)

I 10c

Pokal

C. H. F. Müller, Hamburg, 70er Jahre 19.Jh.;
H. 21,4 cm

Ähnlich I 10b. Abweichend: Kuppa weiter ausladend. Fadeneinlage aus weissen und gelben Einzelfäden. Der hellblaue Schwanz mit einem Faden aus dunklerem Blau konturiert. Die aufgesetzten Warzen am Drachenkopf stärker verschmolzen. — Zungenspitze des Drachenkopfes verloren.

Bonn, Rheinisches Landesmuseum.
Inv. Nr. 68, 0582
Erworben aus Slg. W. Bremen 1968
Lit.: v. Saldern (vgl. I 9a), S.72, Abb. 7 (Nr. 9)

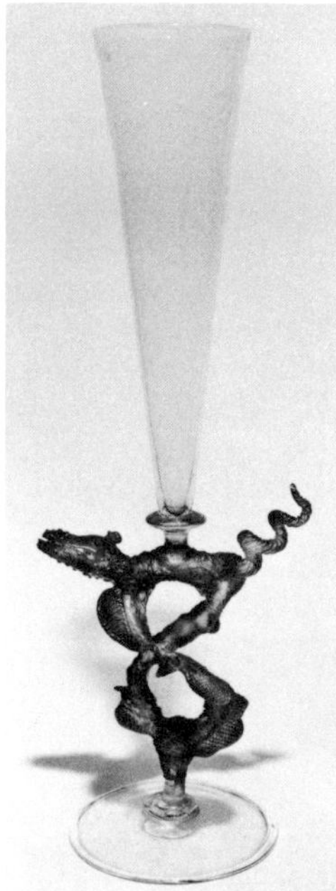

ABB. 17. *In der Art des C.H.F. Müller, vor 1876; Dresden, Kunstgewerbemuseum Schloss Pillnitz (I 10g)*

ABB. 18. *C.H.F. Müller, vor 1874; Nürnberg, Gewerbemuseum, Slg. Völkner (I 10i)*

I 10d

Pokal

C. H. F. Müller?, Hamburg?, 70er Jahre 19.Jh.; H. ca. 26 cm

Sehr ähnlich I 10c. Fadeneinlagen rot, gelb, weiss, Ansätze blau.

Ehm. Slg. Conzen, Düsseldorf. Überliefert als Zeichnung in einem Skizzenbuch Oskar Rauters von 1880 im Kunstmuseum Düsseldorf. Verbleib unbekannt.

I 10e

Pokal

C. H. F. Müller, Hamburg, 70er Jahre 19.Jh.; H. 23 cm

Kleinere Version von I 10b. Das Schaftmotiv gleichmässiger geformt. Teilweise miteinander verschmolzene Fadeneinlagen in der Abfolge blau-gelb-rot-gelb-weiss. Der stark beschädigte Drachenkopf ohne aufgeschmolzene Nuppen.

Lüttich, Musée du Verre. Inv. Nr. B. 700

Erworben auf der Auktion Adelmann, Würzburg 1888

Lit.: v. Saldern (vgl. I 9a), Abb. 8 (Nr. 10)

I 10f
Abb. 16

Deckelpokal

C. H. F. Müller oder Nachahmer, 70er Jahre 19.Jh.; H. 47,6 cm

Sehr klares farbloses Glas. Form von Kuppa und Deckel ähnlich I 1g. Leicht gewölbter Fuss ohne Verdickung am Schaftansatz. Auf der Unterseite abrissartige Aufrauhung, das Glas jedoch in allen Teilen vor der Lampe gearbeitet. Fäden in einem Stab des Schafteinsatzes gelb-blau, im anderen blau-weiss-rot-gelb, im Deckel nur blau-weiss-rot. Ansätze der Schaftverzierung blau, Kopf und Schwanz des Drachen grün. In der Deckelbekrönung Ansätze und Drachen blau.

London, Victoria & Albert Museum. Slg. Buckley, Nr. 323

Lit.: W. Buckley, *The Art of Glass*, London 1939, S.265 und Taf.108; v. Saldern (vgl. I 9a), S.70 f., Abb. 5 (Nr. 6)

I 10g
Abb. 17

Pokal

In der Art des C. H. F. Müller, Niederlande?, vor 1876; H. 23,2 cm

Formal ähnlich I 10a, b. Handwerklich schlecht bewältigt (vgl. I 5c). Die gelb-weissen Einlagen des Schafteinsatzes teilweise zerflossen. Der Einsatz an zahlreichen Stellen wegen ungeeigneten Materials oder unzureichender Kühlung gebrochen. Teile der Stäbe und der blauen Ansätze verloren. Kopf und Schwanz des Drachens gebrochen.

Dresden, Kunstgewerbemuseum Schloss Pillnitz. Inv. Nr. 3355
Erworben 1876 von dem Amsterdamer Händler Goudstikker

I 10h

Pokal

In der Art des C. H. F. Müller, Niederlande?, 70er Jahre 19.Jh.; H. ca. 23 cm

Ähnlich I 10g. Der Schafteinsatz noch stärker beschädigt. Nahezu alle Ansätze einschliesslich Drachenkopf und -schwanz verloren.

Erfurt, Museum für Thüringische Volkskunde. Inv. Nr. VM 440

I 10i
Abb. 18

Flöte

C. H. F. Müller, Hamburg, vor 1874; H. 31,6 cm

Variante von I 10 a-c. Anstelle der Kuppa hoher Flötenkelch. Stäbe des Schafteinsatzes gelblich, Fäden blau-weiss-gelb-rot. Ansätze blau mit feinem Waffelmuster. Die Schaftverzierung mehrfach gebrochen und stark ergänzt.

Nürnberg, Gewerbemuseum der Landesgewerbeanstalt Bayern. Inv. Nr. 2960
Erworben 1874 aus der Slg. Völkner, Prag.

I 10k

Flöte

C. H. F. Müller, Hamburg, 70er Jahre 19. Jh.; H. 31,8 cm

Ähnlich I 10i. Fadeneinlagen gelb-weiss. Der Ansatz der Flöte weniger gut gelungen. Keine Verdickung des Schaftansatzes am Fuss. — Schwanz des Drachen verloren.

Prag, Umĕleckoprůmislové muzeum.
Inv. Nr. 9.646.
Schenkung Lanna, 1906
Lit.: v. Saldern (vgl. I 9a), S. 74, Abb. 11 (Nr. 14)

I 11a
Abb. 19

Flöte

C. H. F. Müller, Hamburg, vor 1874; H. 33 cm

Form ähnlich I 10a, b. Asymmetrisch angelegte Schaftverzierung: gewundene Schlange mit bizarren blauen Ansätzen. Fadeneinlagen rot-weiss-gelb. Der Anschluss des Flötenkelches ähnlich I 10a. — Schafteinsatz mehrfach gebrochen und geklebt.

Nürnberg, Gewerbemuseum der Landesgewerbeanstalt Bayern. Inv. Nr. 2962
Erworben 1874 aus der Slg. Völkner, Prag.

I 11b

Flöte

C. H. F. Müller, Hamburg, 70er Jahre 19. Jh.; H. 31,8 cm

Form und Schafteinsatz ähnlich I 11a. Abweichend: Glasmasse gelblich, kein ausgeprägter Schlangenkopf, Rückenflügel farblos, am Kelchansatz starke Schmauchspuren. — Schaft in der Mitte gebrochen und neu verschmolzen.

ABB. 19. *C. H. F. Müller, vor 1874; Nürnberg, Gewerbemuseum, Slg. Völkner (I 11a)*

ABB. 20. *C. H. F. Müller, vor 1874; Nürnberg, Gewerbemuseum, Slg. Völkner (I 12a, b)*

Kassel, Hessisches Landesmuseum. Inv. Nr. 1966/8
1966 aus dem Kunsthandel erworben

I 11c

Flöte

C. H. F. Müller, Hamburg, 70er Jahre 19. Jh.; H. 31,9 cm

Ähnlich I 11a, b. Fäden gelb-weiss-gelb. Der blaue Rückenkamm mit Parallelriefen gemustert. — Am Flötenansatz gebrochen, Teile des Verbindungsstücks verloren.

Nürnberg, Germanisches Nationalmuseum.
Inv. Nr. Gl 69
Ende des 19. Jh. aus dem Kunsthandel erworben

I 12a
Abb. 20

Pokal

C. H. F. Müller, Hamburg, vor 1874; H. 29,3 cm

Kuppa und Fuss ähnlich I 1i. Schafteinsatz ringförmig, stark tordiert an Vorder- und Rückseite je

ABB. 21. *C.H.F. Müller, vor 1867; Wien, Österreichisches Museum für angewandte Kunst (I 13a)*

ABB. 22. *C.H.F. Müller?, um 1865/70; Leipzig, Museum des Kunsthandwerks, Grassimuseum (I 14b)*

eine sechsblättrige Blüte. Glas des Ringes und der Blüte milchig gelblich, Fadeneinlagen gelb-weiss. Ansätze blau. — Einige Blätter beschädigt und geklebt, desgl. der Ansatz des oberen Schaftendes am Ring. Geschwungene Kämme verloren (falls ehm. vorhanden).

Nürnberg, Gewerbemuseum der Landesgewerbeanstalt Bayern. Inv. Nr. 2963
Erworben 1874 aus der Slg. Völkner, Prag

I 12b
Abb. 20

Pokal

C. H. F. Müller, Hamburg, vor 1874; H. 24,2 cm

Ähnlich I 12a. Keine Verdickung des Schaftansatzes am Fuss. Fäden des Ringes weiss-rot-rosa-türkis, die der Blütenblätter weiss-rot-rosa. — Kämme und Blütenblätter mehrfach gebrochen und geklebt.

Nürnberg, Gewerbemuseum der Landesgewerbeanstalt Bayern. Inv. Nr. 2959
Erworben 1874 aus der Slg. Völkner, Prag

I 12c

Fragment einer Deckelbekrönung

C. H. F. Müller, Hamburg, vor 1874; DM. 8,5 cm

Stark tordierter Ring mit weiss-gelben Fäden und blauen Ansätzen. Die Fäden der Blütenblätter blau-gelb-weiss-rot. Als Bekrönung ein Vogel. — Mehrfach gebrochen und geklebt. Mehrere Ansätze und Teile des Vogels verloren. — Wahrscheinlich nicht zu 12 a oder b gehörig.

Nürnberg, Gewerbemuseum der Landesgewerbeanstalt Bayern. Inv. Nr. 2961/2
Erworben 1874 aus der Slg. Völkner, Prag

I 13a
Abb. 21

Pokal

C. H. F. Müller, Hamburg, 1865/67; H. 28,5 cm

Kuppa- und Fussform sehr ähnlich I 2b. Schafteinsatz mit pilzförmiger Innenfigur. Farblos mit alternierend roten und weissen Fäden. Die dreiecksförmigen Ansätze, Kämme und Schnabelfortsätze blaugrün mit Parallelriefen. — Die Schaftverzierung zweifach geklebt.

Wien, Österreichisches Museum für angewandte Kunst. Inv. Nr. Gl 293/6902
Erworben 1867

Lit.: J. Schlosser, Das alte Glas, Braunschweig 1977, S. 128, Abb. 105

I 13b

Pokal

C. H. F. Müller, Hamburg, um 1865/70; H. 30,3 cm

Sehr ähnlich I 13a. Fäden des Schafteinsatzes weiss-gelb-weiss. Fussrand gewulstet. — Etwa ein Drittel des Fusstellers verloren.

Den Haag, Gemeentemuseum. Inv. Nr. OG–99–54
Legat W. J. H. Mulier 1954
Lit.: Cat. Den Haag (vgl. 1 1h), Nr. 20, S. 26 mit Abb.

I 13c

C. H. F. Müller, Hamburg, 1865/70; H. ca. 31 cm

Ähnlich I 13a. Farben wie I 13b.

Worms, Stiftung Kunsthaus Heylshof. Inv. Nr. 794

I 14a

Balusterpokal

C. H. F. Müller?, Hamburg?, um 1865/67;
H. 39,1 cm

Farblose klare Glasmasse. Steilwandige Kuppa. Im Schaft drei Hohlpuffen mit kleinen Scheiben voneinander abgesetzt. Unter der Kuppa und über dem Fuss je zwei Scheiben. — In zwei Teile zerbrochen.

Wien, Österreichisches Museum für angewandte Kunst. Inv. Nr. Gl 294/6903
Erworben 1867 gemeinsam mit I 13a.

Lit.: Das Kunsthandwerk (vgl. I 9b), Jg.I, 1874, Tf.19

I 14b
Abb. 22

Balusterpokal

C. H. F. Müller?, Hamburg?, um 1865/70;
H. 31,7 cm

Sehr ähnlich I 14a. Form der Kuppa im unteren Teil stärker gerundet.

Leipzig, Museum des Kunsthandwerks-Grassi Museum. Inv. Nr. 08.4
Erworben aus Slg. Spitzer, Paris; Auktion 1908

I 14c

Balusterpokal

C. H. F. Müller?, Hamburg?, um 1865/70;
H. 32,5 cm

Variante von I 14 a/b. Statt der drei Hohlpuffen nur zwei, diese jedoch grösser gebildet.

Halle, Staatliche Galerie Moritzburg. Inv. Nr. 374
Erworben 1914 im Berliner Kunsthandel

Lit.: Fritz Kämpfer, *Viertausend Jahre Glas*, München 1966; Abb. 135, u. S.282.

I 15a
Abb. 23

Pokal

C. H. F. Müller?, Hamburg?, 70er Jahre 19.Jh.;
H. 24 cm

Kuppa- und Fussform sehr ähnlich I 3d u.a. Farblose, klare, leicht blasige Glasmasse. Im Schaft ein krokodilartiges Ungeheuer mit aufgesperrtem Rachen, Kopf und Hals in der Art der Drachen von I 10 mit farblosen kleinen Nuppen besetzt. — Schaft im oberen Teil gebrochen und neu verschmolzen, am Fussansatz gebrochen und geklebt.

Frankfurt, Museum für Kunsthandwerk.
Inv. Nr. 7013
Schenkung Kessler-Kollings 1934

Lit.: Kat. Frankfurt (vgl. I 11), Nr. 139, S.72 mit Abb.

ABB. 23. *C. H. F. Müller?, 70er Jahre 19.Jh.; Frankfurt, Museum für Kunsthandwerk, Schenkung Kessler-Kollings (I 15a)*

ABB. 24. *C. H. F. Müller?, 70er Jahre 19.Jh.; Slg. Peill & Putzler, Düren (I 15b)*

I 15b
Abb. 24

Pokal

C. H. F. Müller?, Hamburg?, 70er Jahre 19.Jh.;
H. 24,5 cm

Kuppa- und Fussform ähnlich I 15a. Schaftansatz weniger stark gestaucht. Farbloses Glas. Im Schaft aufgerichteter Ziegenbock, am ganzen Körper mit feinen hellblauen Nuppen besetzt. — Am Körper des Bockes kleinere Sprünge.

Düren, Slg. Peill und Putzler

Lit.: Ausst. Kat. *Meisterwerke der Glaskunst aus internationalem Privatbesitz*, Düsseldorf 1968, Nr. 103, S.41

I 15c

C. H. F. Müller?, Hamburg?, 70er Jahre 19.Jh.;
H. 25,5 cm

Sehr ähnlich I 15b.

Hannover, Kestner Museum, Inv. Nr. 1920,8

Erworben 1920 aus Privatbesitz

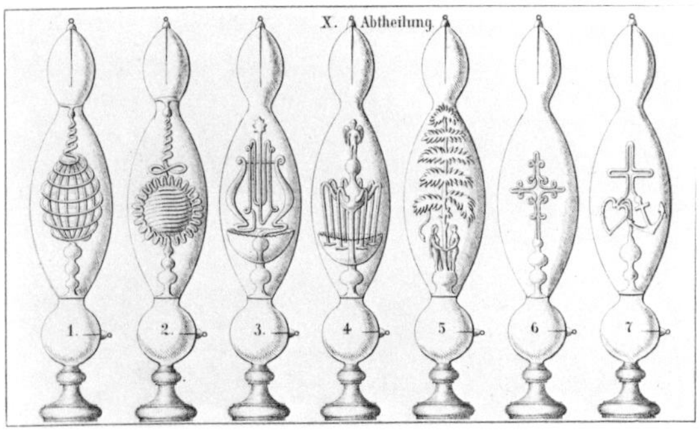

ABB. 25. *C.H.F. Müller, Elektrische Gasentladungsröhren aus einem* Preis-Verzeichnis *von 1880*

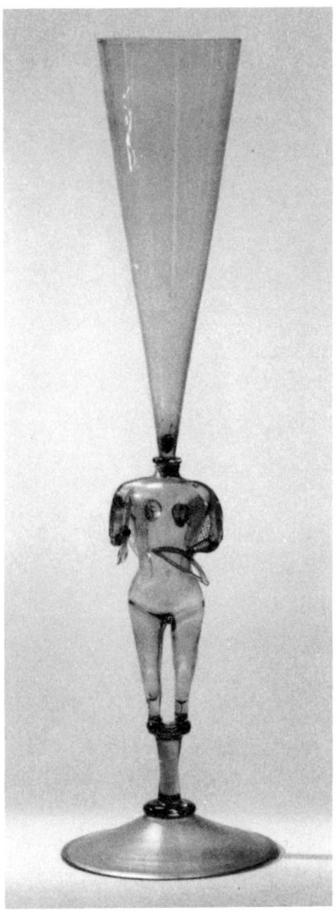

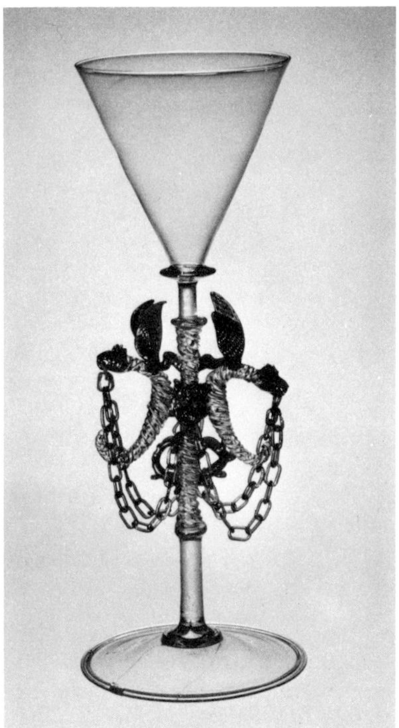

ABB. 26. *C.H.F. Müller, 70er Jahre 19.Jh.: Stockholm, Nationalmuseum, Slg. Bendix (I 16b)*

ABB. 27. *C.H.F. Müller, 70er Jahre 19.Jh.: Jerome Strauss, State College, Pa. (I 17)*

I 15d

C. H. F. Müller?, Hamburg?, 70er Jahre 19. Jh.;
H. 28,5 cm

Variante des Typs I 15b, c. Der Ziegenbock kräftiger gebildet, auf einer T-förmigen Stütze aufstehend. Keine farbigen Nuppen. Merfach gesprungen und geklebt.

Hannover, Kestner Museum, Inv. Nr. 1920,7

Erworben 1920 aus Privatbesitz; gleiche Provenienz wie I 15c

***I 16a**
Abb. 54, Nr. 14

Pokal

C. H. F. Müller, Hamburg, 70er Jahre 19. Jh.;
H. 25,3 cm

Als Schafteinsatz Amorfigürchen mit Pfeil und Bogen. Leicht braunstichiges Glas. Abriss unter dem Fuss. Mehrfach gebrochen und geklebt. Kuppa neu geblasen und angeklebt. Für den ursprünglichen Zustand vgl. Abb. 54, Nr. 14.

Lüttich, Musée du Verre. Inv. Nr. B 1248
Slg. Baar

Lit.: Chambon (s.Anm.40), pl. XXII, Abb. 67 (heutiger Zustand).

I 16b
Abb. 26

Flöte

C. H. F. Müller, Hamburg, 70er Jahre 19. Jh.;
H. 36,5 cm

Figürchen ähnlich I 16a. Gelbliches Glas mit Grau-stich, sehr ähnlich II 7. Pfeil, Bogen und Flügel blau. Kopf des Amor verloren, der Flötenkelch im Halsstumpf verklebt.

Stockholm, Nationalmuseum. Inv. Nr. NM 16/1902
Schenkung W. Bendix 1902 (seit 1887 im Museum deponiert)

I 17
Abb. 27

Pokal

C. H. F. Müller, Hamburg, 70er Jahre 19. Jh.;
H. 25,7 cm

Spitzkuppa sehr ähnlich I 8. Schaft aus stark tordiertem Fadenglasstab mit Einlagen in weiss-gelb-rot-blau, flankiert von zwei nach auswärts geschwungenen geflügelten Drachen mit je zwei Ketten im Maul. Ketten und Ansätze grün. — Zwei Flügel in Kunststoff ergänzt. Fuss und Schaft gebrochen und geklebt, geringe Teile der Ansätze verloren.

Corning, N.Y., The Corning Museum of Glass.
Inv. Nr. S.1597 (Strauss Collection)

ABB. 28. *a) Schnittdarstellung eines lampengeblasenen Deckels der Gruppe I, b) eines am Ofen gefertigten Deckels, c) eines Deckels der Gruppe III*

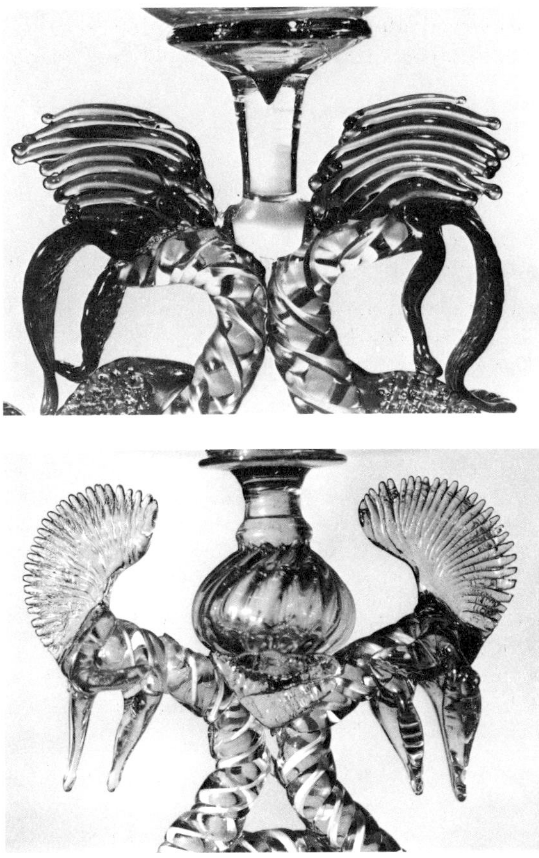

ABB. 29. *a) Zierkämme, Lampenarbeit (Detail Abb. 4), b) Zierkämme, Ofenarbeit. Detail des Pokals Nr. 201 der Slg. Bremen in Landesmuseum Bonn*

ABB. 30. *Werkstatt eines Lampenbläsers. Kupferstichillustration aus Joh. Kunckels* Ars Vitraria, *1679*

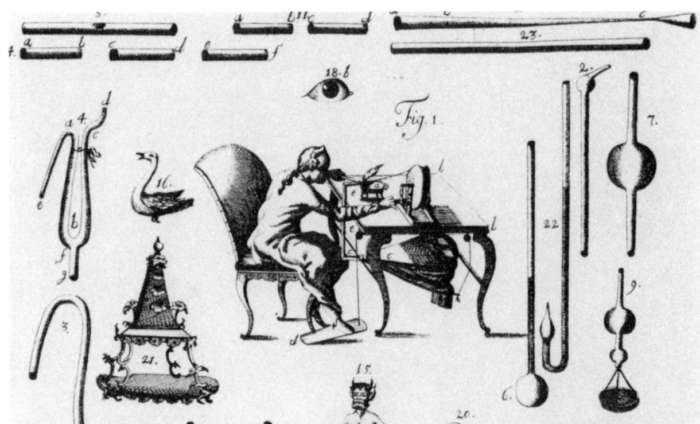

ABB. 31. *Glasspinner bei der Arbeit, Werkstücke. Kupferstichillustration aus* Die Glasschmelzkunst bey der Lampe, *Wien 1769*

Gruppe II
Frei gestaltete Gebrauchsgläser in venezianischem Stil

***II 1a**
Abb. 32/1

Flöte

Entwurf Fr. Keller-Leuzinger, Ausführung C. H. F. Müller, Hamburg, 1877/79; H. 14,7 cm

Leicht hornfarbenes Glas, die drei Hohlkugeln des Balusterschafts besetzt mit roten, blauen und gelbgrünen Flachnuppen, darauf kleine farblose und grüne Perlnuppen. Grüne Perlnuppen auch auf den beiden Scheibchen, die Kelch und Fuss vom Schaft trennen.

Hamburg, Museum für Kunst und Gewerbe.
Inv. Nr. 1879.252
Von Fr. Keller-Leuzinger 1879 für 6, — Mark erworben.

Lit.: Ausst. Kat. Historismus Hamburg (s.Anm.7), Nr. 328, S.198 mit Abb.

***II 1b**

Champagnerglas

Entwurf verm. Fr. Keller-Leuzinger, Ausführung C. H. F. Müller, Hamburg, um 1877/79; H. 25,5 cm

Kelch aus röthlichem Glas, die drei Knäufe des Stengels mit bunten Glaskügelchen besetzt (Inventareintrag).

Hamburg, Museum für Kunst und Gewerbe,
Inv. Nr. 1893.67

1893 von C. H. F. Müller für 9, — Mark erworben. Laut Inventarvermerk *total zerbrochen*. Die Fragmente verschollen.

***II 2**
Abb. 32/2

Fussbecher

Entwurf, Ausführung und Datierung wie II 1a;
H. 11 cm

Kuppa zu ovalem Querschnitt verformt. Fuss mit spiralig aufgesponnenem Faden. Dunkelgrün mit blaugrünen Auflagen.

Flensburg, Städtisches Museum. Inv. Nr. 307

Erworben aus einer Gruppe von Gläsern, die das Hamburger Museum für Kunst und Gewerbe 1879 von Fr. Keller-Leuzinger ankaufte (ehm. Inv. Nr. 1879.255; vgl. II 1a)

***II 3**
Abb. 32/2

Römer

Entwurf, Ausführung und Datierung wie II 1a;
H. 10,5 cm

Glasmasse, Farbe der Auflagen und Bildung des Fusses wie II 2.

Flensburg, Städtisches Museum. Inv. Nr. 308
Provenienz wie II 2. Ehm. Inv. Nr. Hamburg 1879.256

***II 4**
Abb. 32/2

Römer

Entwurf, Ausführung und Datierung wie II 1a;
H. 9,3 cm

Glasmasse, Farbe der Auflagen und Bildung des Fusses wie II 2.

Flensburg, Städtisches Museum. Inv. Nr. 309
Provenienz wie II 2. Ehm. Inv. Nr. Hamburg 1879.257

II 5
Abb. 33

Römer

Entwurf Fr. Keller-Leuzinger?, Ausführung C. H. F. Müller zugeschr. Hamburg, um 1880; H. 11,8 cm

Gelblich mit Graustich; Flachnuppen mit grünen Perlnuppen. Stockholm, Nationalmuseum. Inv. Nr. NM X 36

***II 6**
Abb. 34/1

Kelchglas

Entwurf R. Bichweiler, Ausführung C. H. F. Müller zugeschr. Hamburg, 1880; H. 10,5 cm

Kelch leicht hornfarben, Fuss und Baluster fast farblos. Der Schaft besetzt mit gelbgrünen Flachnuppen und Perlnuppen in blau, grün, rot; grüne Perlnuppen auf den beiden Scheibchen zwischen Kelch, Schaft und Fuss.

Hamburg, Museum für Kunst und Gewerbe.
Inv. Nr. 1880.386
Von der Kunstgewerblichen Werkstatt R. Bichweiler in Hamburg 1880 für 4,50 Mark erworben.

Lit.: Ausst. Kat. Historismus Hamburg (s.Anm.7), Nr. 329, S.199 mit Abb.

32/1

32/2

ABB. 32/1. *Entwurf Fr. Keller-Leuzinger, Ausführung C. H. F. Müller, 1877/79; Hamburg, Museum für Kunst und Gewerbe. Ankauf vom Entwerfer (II 1a)*

ABB. 32/2. *Entwurf Fr. Keller-Leuzinger, Ausführung C. H. F. Müller, 1877/79; Flensburg, Städtisches Museum (II 4, 3, 2)*

33

34/1

34/2

ABB. 33. *Entwurf Fr. Keller-Leuzinger?, Ausführung C. H. F. Müller zugeschr., um 1880; Stockholm, Nationalmuseum, Slg. Bendix (II 5)*

ABB. 34/1. *Entwurf R. Bichweiler, Ausführung C. H. F. Müller zugeschr., 1880; Hamburg, Museum für Kunst und Gewerbe, Ankauf vom Entwerfer (II 6)*

ABB. 34/2. *Entwurf Fr. Keller-Leuzinger oder R. Bichweiler, Ausführung C. H. F. Müller zugeschr., um 1877/80; The Corning Museum of Glass (II 7b)*

ABB. 35. *Entwurf R. Bichweiler, Ausführung C. H. F. Müller zugeschr., 1880; Hamburg, Museum für Kunst und Gewerbe, Ankauf vom Entwerfer (II 8a, b)*

ABB. 36. *Zeichnungen nach von R. Bichweiler entworfenen Gläsern. Aus einem Skizzenbuch Oskar Rauters von 1881 im Kunstmuseum Düsseldorf (II 9, 10)*

ABB. 37. *Entwurf R. Bichweiler, Ausführung C. H. F. Müller zugeschr., um 1880; Stockholm, Nationalmuseum, Slg. Bendix (II 9)*

ABB. 38. *Entwurf R. Bichweiler, Ausführung C. H. F. Müller zugeschr., um 1880; Düsseldorf, Kunstmuseum (II 10)*

ABB. 39. *Entwurf R. Bichweiler, Ausführung C. H. F. Müller zugeschr., um 1880; Slg. J. & L. Lobmeyr, Wien (II 11)*

II 7a

Flöte

Entwurf Fr. Keller-Leuzinger oder R. Bichweiler, Ausführung C. H. F. Müller zugeschr. Hamburg, um 1877/79; H. 25 cm

Im Aufbau ähnlich der Flöte II 1a. Schaft ähnlich II 3. Der Kelch rosé mit violettem Stich (vgl. II 1b), der gerippte Balusterschaft und der nur am Rand leicht gerippte Fuss hornfarben. Auf dem Baluster mit Perlnuppen besetzte Flachnuppen in blau, gelb und rot, die mittlere Reihe durch girlandenartig angeordnete Perlnuppen verbunden.

Eisenach, DDR. Thüringer Museum, Schloss am Markt. Inv. Nr. B 5/487

II 7b
Abb. 34/2

Flöte

Entwurf Fr. Keller-Leuzinger oder R. Bichweiler, Ausführung C. H. F. Müller zugeschr. Hamburg, um 1877/80; H. 26,2 cm

Aufbau wie II 7a. Farbloses Glas mit mehrfarbigen Auflagen.

Corning, N.Y., The Corning Museum of Glass. Inv. Nr. 77.3.38

***II 8a, b**
Abb. 35

Zwei Weingläser

Entwurf R. Bichweiler, Ausführung C. H. F. Müller zugeschr. Hamburg, 1880; H. 12 (a) und 14,2 (b) cm

Austeigender Rippenfuss, darüber Baluster und Hohlscheibe mit angeblasenem Kugelkranz. Farblos, leicht hornfarben.

Hamburg, Museum für Kunst und Gewerbe. Inv. Nr. 1880.387 (a); o.Nr.(b).
Das kleinere Glas von der Kunstgewerblichen Werkstatt R. Bichweiler in Hamburg 1880 für 4,— Mark erworben.

Lit.: Ausst. Kat. Historismus Hamburg (s.Anm.7), Nr. 330, S. 199 mit Abb.

***II 9**
Abb. 37

Kelchglas

Entwurf R. Bichweiler, Ausführung C. H. F. Müller zugeschr. Hamburg, um 1880; H. 14,5 cm

Fuss und Schaft ähnlich II 5a, b. Hornfarbenes Glas.

Stockholm, Nationalmuseum. Inv. Nr. NM 22/1902
Schenkung W. Bendix 1902 (seit 1887 im Museum deponiert)

***II 10**
Abb. 38

Flöte

Zuschreibung und Datierung wie II 6; H. 37,7 cm

Im Schaft eine leicht gedrückte rechtsläufig gerippte Kugel eingefasst von je einer Hohlscheide mit angeblasenem Kugelkranz. Farblos mit starkem Olivstich. Schaft gebrocken und geklebt

Düsseldorf, Kunstmuseum. Inv. Nr. D 1237
1887 erworben

II 11
Abb. 39

Pokal

Zuschreibung und Datierung wie II 6; H. 21 cm

Fuss und Schaft wie II 7. Bräunliches Glas.
Wien, Slg. J. & L. Lobmeyr

***II 12**
Abb. 40

Senftopf mit Löffel

Entwurf R. Bichweiler, Ausführung C. H. F. Müller zugeschr.
Hamburg, 1880; H. 6,2; DM. 7 cm. Löffel 10 cm lang.

Grünliches Glas, braune Flachnuppen mit farblosen und grünen Perlnuppen, die Nuppenreihen an Hals- und Fussansatz grün. Der Löffel gelblich mit gedrehter Rippung und farbloser Kelle; vielfarbige Perlnuppen am Griff. — Deckel verloren.

Hamburg, Museum für Kunst und Gewerbe.
Inv. Nr. 1880.389.
1880 für 6, — Mark in der Kunstgewerblichen Werkstatt Robert Bichweiler erworben.

Lit.: Ausst. Kat. Historismus Hamburg (s.Anm.7), Nr. 331, S.200 mit Abb.

II 13a

Flöte

Entwurf R. Bichweiler oder Fr. Keller-Leuzinger? Ausführung C. H. F. Müller zugeschr.

Hamburg, um 1880; H. 40,2 cm

Hohe schlanke Flöte aus bräunlichem Glas auf spiralig geripptem, aus 5 Hohlkugeln gebildeten

ABB. 40. *Entwurf R. Bichweiler, Ausführung C. H. F. Müller zugeschr., 1880; Hamburg, Museum für Kunst und Gewerbe (II 12)*

Schaft. Fuss mit linksläufig gewirbelter Rippung und gewulstetem Rand.

Amsterdam, Rijksmuseum. Inv. Nr. 4694
1881 in Amsterdam erworben.

II 13b

Wie II 10a; H. 55,5 cm

Amsterdam, Rijksmuseum. Inv. Nr. NM 10754:35
Legat A. J. Enschedé 1896

II 13c

Wie II 10a; H. 53,7 cm

Arnhem, Gemeentemuseum. Inv. Nr. VS 186
Leihgabe aus Privatbesitz

II 13d

Wie II 10a; H. 48,1 cm

Den Haag, Gemeentemuseum. Inv. Nr. OG–204–54
Legat W. J. H. Mulier 1954

Lit.: Cat. Den Haag (vgl. I 1h), S.26, Nr. 29 mit Abb.

II 13e

Wie II 10a; H. 58,3 cm

Den Haag, Gemeentemuseum. Inv. Nr. OG 1–1953
1953 durch Schenkung erworben

Lit.: Cat. Den Haag (vgl. I 1h), S.29, Nr. 24 mit Abb.

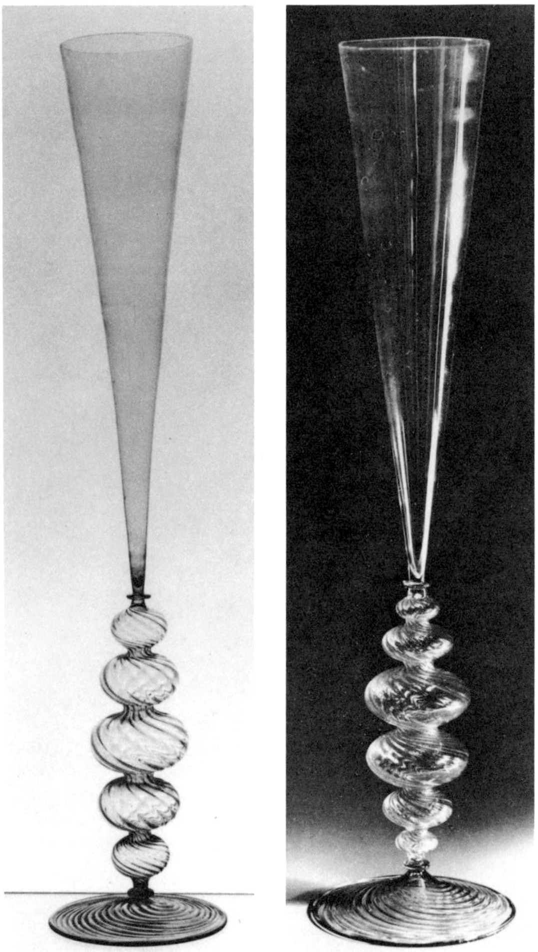

A b b. 41. *Entwurf R. Bichweiler?, Ausführung C. H. F. Müller zugeschr., um 1880; a) Oslo, Kunstindustriemuseet (II 13h), b) Hannover, Kestner-Museum (II 13n)*

II 13f, g

Wie II 10a; H. 59,3; 57,1 cm

Kopenhagen, Kunstindustriemuseet.
Inv. Nr. A 33/1940
1940 aus Slg. Oluf Thomsen

II 13h
Abb. 41a

Wie II 10a; H. 59 cm

Oslo, Kunstindustriemuseet. Inv. Nr. OK 4603
1894 erworben

II 13i

Wie II 10a; H. 50,5 cm

Wien, Österreichisches Museum für angewandte Kunst. Inv. Nr. Gl 2451/27.162
1927 erworben

II 13k

Wie II 10a; H. 56 cm

Stockholm, Nationalmuseum. Inv. Nr. NM 15/1902
1902 Schenkung Bendix; seit 1887 im Museum

II 13l, m

Ähnlich II 10a; H. ca. 50 cm; Form der Balusterkugeln etwas abweichend.
Ähnlich II 10a; H. ca. 50 cm. Sechs Balusterkugeln im Schaft.

Buenos Aires, Slg. Ernesto Wolf

Lit.: Ausst. Kat. *Glas aus vier Jahrtausenden*, Zürich 1956, Nr. 25 mit Abb.

II 13n
Abb. 41b

Wie II 10a; H. 44,4 cm. Sechs Balusterkugeln im Schaft.
Hannover, Kestner-Museum. Inv. Nr. 1966, 103

II 13o

Ähnlich II 10a; H. 50,8 cm. Vier Balusterkugeln im Schaft, vermutlich ehemals fünf; die fünfte und der obere Schaftring verloren, die Bruchstelle neu verschmolzen. Fuss gebrochen und geklebt.

Corning, N.Y., The Corning Museum of Glass.
Inv. Nr. 56.3.69

Gruppe III

Frei gestaltete Ziergläser in venezianischem Stil

***III 1a**
Abb. 42

Deckelpokal

C. H. F. Müller, Hamburg, um 1875/80; H. 36 cm

Konische Kuppa mit gerundetem Ansatz. Dünner massiver Stengelschaft, am Kuppaansatz zu einem Wulst gestaucht, desgleichen oberhalb des flachen, am Rand leicht verdickten Fusstellers. Farbloses Glas, leicht hornfarben. Einsatz in umgekehrter Herzform mit verschlungener Innenfigur ähnlich I 7, der

ABB. 42. *C. H. F. Müller, um 1875/80; Hamburg, Museum für Kunst und Gewerbe (III 1)*

ABB. 43. *Ausführung C. H. F. Müller zugeschr., um 1880; The Corning Museum of Glass (III 2)*

ABB. 44. *Ausführung C. H. F. Müller zugeschr., um 1880; Frankfurt, Museum für Kunsthandwerk (III 3a)*

Stab jedoch eng zusammengedreht. Fadeneinlagen rot und gelb mit wenig blau. Ansätze am Pokal blaugrün, am Deckel blau. Deckel mit geschwungenem Profil und überkragendem Rand. Verzierung ähnlich dem Schafteinsatz, als Bekrönung ein Phantasievogel mit nixenartig geteiltem Schwanz. — Die Kämme der Schaftverzierung und ein "Schnabel" der Deckelbekrönung verloren.

Hamburg, Museum für Kunst und Gewerbe.
Inv. Nr. 1915.1
1915 aus Hamburger Privatbesitz erworben.

Lit.: Ausst. Kat. Historismus Hamburg (s.Anm.7), Nr. 327, S.197 mit Abb.

III 1b

Deckel

Entwurf Fr. Keller-Leuzinger oder R. Bichweiler? Ausführung C. H. F. Müller zugeschr.

Hamburg, Ende 70er/Anfang 80er Jahre 19.Jh.; H. 14,1 cm

Ähnlich III 1a. Ausführung sorgfältiger und detaillierter. Verbindungsstück zwischen Deckelkalotte und Aufsatz kürzer. Fadeneinlagen gelb-weiss-rot-blau. — Bekrönung der Deckelverzierung (Vogel?) und die oberen Schnabelfortsätze verloren.

Lüttich, Musée du Verre. Inv. Nr. B.563
Slg. Baar; erworben 1885 in Lüttich, zusammen mit I 2b, jedoch nicht zugehörig.

III 2
Abb. 43

Deckelpokal

Zuschreibung und Datierung wie III 1b; H. 40,3 cm

Hornfarbenes Glas. Der Schafteinsatz Variante des Typs I 1, gebildet aus zwei Stäben. Fadeneinlagen weiss-rot-gelb-blau. Grüne Ansätze. — Kämme und Schnabelfortsätze der Schaftverzierung verloren.

Corning, N.Y., The Corning Museum of Glass.
Inv. Nr. 59.3.47a, b

III 3a
Abb. 44

Deckelpokal

Zuschreibung und Datierung wie III 1b; H. 36,5 cm

Kuppa, Deckel und Fussform wie III 2. Glas gelblich. Schafteinsatz entsprechend I 5, jedoch gleichmässiger gearbeitet. Fadeneinlagen bandartig violett-rot-weiss-gelb, leicht tordiert, Ansätze blaugrün. Die unteren Ansätze mit gröberem, die Kämme und die oberen Ansätze mit feinerem Waffelmuster. Deckelaufsatz entsprechend der Schaftverzierung, als Bekrönung ein gedrehtes Türmchen. Ansätze mit feinem Waffelmuster.

Frankfurt, Museum für Kunsthandwerk.
Inv. Nr. 12147
1951 durch Schenkung erworben.
Lit.: Kat. Frankfurt (vgl. I 1c), Nr. 166, S.81 mit Abb.

III 3b

Deckelpokal

Zuschreibung und Datierung wie III 1b; H. 38 cm

Ähnlich III 3a. Abweichend: der Fussrand leicht gewellt, Stab des Schafteinsatzes stark tordiert, Fadeneinlagen rot-gelb-weiss. Als Deckelbekrönung zwei symmetrisch angeordnete Kämme. Alle gekniffenen Partien mit feinem Waffelmuster. — Eine der unteren Schlingen des Deckels mit den zugehörigen Ansätzen verloren.

Bremen, Landesmuseum für Kunst- und Kulturgeschichte. Focke Museum. Ohne Nr.

III 4a

Deckelpokal

Zuschreibung und Datierung wie III 1b; H. 38,7 cm

Kuppa- und Fussform ähnlich III 3a. Baluster rechtsläufig gerippt, Deckel etwas gedrückter. Schafteinsatz räumlich doppelläufig ähnlich I 9. Deckelverzierung analog geformt, als Bekrönung ein Vogel, flankiert von ungeschwungenen Kämmen. Hornfarbenes Glas. Fadeneinlagen: Schaft weiss-rot-gelb, Deckel gelb-weiss-rot-blau-weiss. Alle Ansätze blau. — Die Deckelkalotte gesprungen.

Stockholm, Nationalmuseum. Inv. Nr. 14/1902
Schenkung W. Bendix 1902 (seit 1887 im Museum deponiert)

Lit.: v. Saldern (vgl. I 9a), S.68 ff., Abb. 3 (Nr. 3)

III 4b
Abb. 45

Deckelpokal

Zuschreibung und Datierung wie III 1b; H. 37,4 cm

Leicht vereinfachte Variante von II 4a. Abweichend: Kämme ohne innere Schwingung, Verzicht auf schnabelartige Fortsätze. Fadeneinlagen: Schaft blau-rot-weiss-gelb, Deckel blau-rot-weiss. Ansätze grün.

Stockholm, Nationalmuseum. Inv. Nr. NM 13/1902
Schenkung W. Bendix 1902 (seit 1887 im Museum deponiert)

Lit.: v. Saldern (vgl. I 9a), S.68, Nr. 4 ohne Abb.

III 4c, d

Pokale

Zuschreibung und Datierung wie III 1b; H. 21,5 und 27,5 cm

Ähnlich III 4a. Gelbliches Glas. Schafteinsatz mit gelbem Faden und türkisblauen Ansätzen.

State College, Pennsylvania. Jerome Strauss Collection. Nr. S.519

III 5a
Abb. 46

Deckelpokal

Zuschreibung und Datierung wie III 1b; H. 38,3, ohne Deckel 23,3 cm

Fuss-, Schaft- und Deckelform und der allgemeine Aufbau ähnlich II 4a, b. Fadeneinlagen blau-rot-weiss-gelb, weiss kaum wahrnehmbar. Anstelle der Kämme ein Drachenkopf aus massivem Glas, der

A B B. 45. *Ausführung C. H. F. Müller zugeschr., um 1880; Stockholm, Nationalmuseum, Slg. Bendix (III 4b)*

A B B. 46. *Ausführung C. H. F. Müller zugeschr., um 1880; Düsseldorf, Kunstmuseum, (Pokal), Lüttich, Musée du Verre (Deckel) (III 5a)*

A B B. 47. *Ausführung C. H. F. Müller zugeschr., um 1880; The Corning Museum of Glass (III 5b)*

A B B. 48. *Ausführung C. H. F. Müller zugeschr., um 1880; Bremen, Focke-Museum (III 6)*

Rachen besetzt mit zwei Zahnreihen aus weissen Punkten, das Auge weiss mit brauner Pupille. Auf der gegenüberliegenden Seite stark gewundener Schwanz, mit einem gesondert aufgelegten Faden konturiert (vgl. die Gläser des Typs I 10). Die Glasmasse hornfarben, alle Ansätze grün.

Düsseldorf, Kunstmuseum (Pokal). Inv. Nr. 12659 1888 im Lütticher Kunsthandel erworben

Lüttich, Musée du Verre (Deckel). Inv. Nr. B. 926 Slg. Baar; 1890 im Lütticher Kunsthandel erworben.

Lit.: v. Saldern (vgl. I 9a), S.67 ff., Abb. 1 (Nr. 1) und Nr. 7 ohne Abb.

III 5b
Abb. 47

Deckelpokal

Zuschreibung und Datierung wie III 1b; H. 35,6 cm

Ähnlich III 5a. Die Gesamtproportion schlanker. — Mehrere Ansätze der Schaftverzierung verloren. Deckel am Ansatz der Bekrönung gebrochen und geklebt, Drachenkopf und -schwanz verloren.

Corning, N.Y., The Corning Museum of Glass. Inv. Nr. 51.3.115

Lit.: v. Saldern (vgl. I 9a), S.68 ff., Abb. 2 (Nr. 2)

ABB. 51. *Ausführung C. H. F. Müller zugeschr., um 1880; Stockholm, Nationalmuseum, Slg. Bendix (III 9a, b)*

ABB. 49. *Ausführung C. H. F. Müller zugeschr., um 1880; Stockholm, Nationalmuseum, Slg. Bendix (III 7)*

ABB. 50. *Ausführung C. H. F. Müller zugeschr., um 1880; Frankfurt, Museum für Kunsthandwerk (III 8)*

III 6
Abb. 48

Pokal

Zuschreibung und Datierung wie III 1b; H. 27,1 cm

Kuppa-, Schaft- und Fussform wie III 2a. Im Schafteinsatz zwei räumlich gegenläufig geführte, leicht tordierte Stäbe. Kuppaansatz verbunden durch zwei umeinandergewundene Fadenglasstäbe. Einlagen rot-gelb-weiss, die Ansätze blau. — Kleiner Sprung am Kuppaansatz

Bremen, Landesmuseum für Kunst- und Kulturgeschichte. Focke Museum. Ohne Nr.

III 7
Abb. 49

Deckelpokal

Zuschreibung und Datierung wie III 1b; H. 39,5 cm

Kuppa-, Fuss-, Schaft-, Deckelform und Glasmasse wie III 3a. Im Schafteinsatz und als Deckelverzierung Drachen, gewunden in Form einer 8. Die Köpfe ähnlich III 5a, b. Seitlich angesetzte Flügel mit feiner Waffelmusterung. Auf dem Rücken der Fabeltiere bizarr gezackte Kämme ähnlich I 3g und I 6a. Fadeneinlagen gelb-weiss-rot, die Ansätze blau.

Stockholm, Nationalmuseum. Inv. Nr. 12/1902
Schenkung W. Bendix 1902 (seit 1887 im Museum deponiert)

III 8
Abb. 50

Deckelpokal

Zuschreibung und Datierung wie III 1b; H. 34 cm

Fuss- und Deckelform wie III 3a. Die Kuppa etwas höher und steilwandiger, der vertikalgerippte Baluster stärker ausgeblasen, und mit einem Kranz von feinen Nuppen verziert. Schafteinsatz und Deckelbekrönung kugelförmig, gebildet aus Fadenglasseg-

ABB. 52. *Ausführung C. H. F. Müller, um 1880; a) Amsterdam, Rijksmuseum, Legat Enschedé (III 10c), b) Stockholm, Nationalmuseum, Slg. Bendix (III 10g), c) Berlin, Kunstgewerbemuseum Schloss Köpenick (III 10f)*

menten mit Nuppenauflagen. Die Fäden violett-rot-weiss-gelb, die Auflagen im Schaft blau, am Deckel grün. — Bekrönung der Deckelverzierung verloren.

Frankfurt, Museum für Kunsthandwerk.
Inv. Nr. 12149
1951 durch Schenkung erworben

Lit.: Kat. Frankfurt (vgl. I 1i), Nr. 167, S.81 mit Abb.

III 9a, b
Abb. 51

Zwei Pokale

Zuschreibung und Datierung wie III 1b; H. 22 cm

Fuss wie III 3a ff., Schaftform wie II 5, 6. Hornfarben, im Schafteinsatz vierpassartige stark (a) und leicht (b) tordierte Schlaufen mit gelbweissen Fadeneinlagen. Im Zentrum an Vorder- und Rückseite je eine sechsblättrige grüne Blüte mit roter Mitte. Mehrere der Blütenblätter verloren.

Stockholm, Nationalmuseum. Inv. Nr. NM 17/18/1902
Schenkung W. Bendix 1902 (seit 1887 im Museum deponiert)

Lit.: H. Seitz, Glaset förr och nu, Stockholm 1933, Tf. 30

***III 10a**

Doppeldrachen-Glas

Entwurf Fr. Keller-Leuzinger oder R. Bichweiler?
Ausführung C. H. F. Müller

Hamburg, Ende 70er/Anfang 80er Jahre;
H. 16,7 cm

Um den balusterförmigen Stengel winden sich zwei symmetrisch angebrachte Drachen von blauem Glas mit Ansätzen von klarem Glas (Inventareintrag).

Hamburg, Museum für Kunst und Gewerbe.
Inv. Nr. 1893–65
1893 von C. H. F. Müller für 10,— Mark erworben.
Laut Inventarvermerk *total zerbrochen*. Die Fragmente verschollen.

***III 10b**

Doppeldrachen-Glas

Entwurf Fr. Keller-Leuzinger oder R. Bichweiler?
Ausführung C. H. F. Müller

Hamburg, Ende 70er/Anfang 80er Jahre;
H. 21,7 cm

Um den balusterförmigen Stengel winden sich zwei symmetrisch angebrachte Drachen, innen mit farbigen Spiralfäden versehen; aussen mit blaugrünen Ansätzen verziert (Inventareintrag).

Hamburg, Museum für Kunst und Gewerbe.
Inv. Nr. 1893–66
1893 von C. H. F. Müller für 12,— Mark erworben.
Laut Inventarvermerk *total zerbrochen*. Die Fragmente verschollen.

III 10c
Abb. 52a

Pokal

Zuschreibung und Datierung wie III 10a, b;
H. 22,5 cm

Fuss- und Kuppaform ähnlich III 3a. Kuppa im unteren Teil in Wellen gerippt. Schaft als gestreckter Baluster mit Vertikalrippen ausgebildet, darüber abgesetzte Hohlscheibe mit Kranz aus kleinen Halbkugeln. Das kurze zur Kuppa überleitende massive Schaftstück durch ein Scheibchen unterteilt.

Um den Baluster winden sich zwei symmetrisch angeordnete Schlangen, deren Köpfe auf der Hohlscheibe aufruhen. Fadeneinlagen blau-weiss-rot-gelb, die Kämme und Ansätze der Schlangen grün. Einer der Kämme verloren.

Amsterdam, Rijksmuseum. Inv. Nr. 10754:8
Legat A. J. Enschedé 1896

III 10d

Pokal

Zuschreibung und Datierung wie III 10a, b;
H. 22 cm

Ähnlich III 10c. Etwas schlankere und höhere Kuppaform (vgl. III 8).
Linksläufige Rippung der unteren Kuppapartie.
Farbangaben waren nicht zu beschaffen.

Freiburg, Galerie Peter Lankoff (Dez. 1976)
Abb.: *Die Kunst*, 88. Jg. 1976, S. 764 (*Venezianisches Glas. 17. Jh.*)

III 10e

Pokalfragment

Zuschreibung und Datierung wie III 10a, b;
H. 13,5 cm

Form, Aufbau und Farben wie III 10c.
Die Kuppa verloren.

Amsterdam, Slg. C. A. Burgers

III 10f
Abb. 52c

Pokal

Zuschreibung und Datierung wie III 10a, b;
H. 24,7 cm

Allgemeiner Aufbau wie III 10c. Kuppa etwas schlanker und höher (vgl. III 8). In der unteren Partie aufgeschmolzene Vertikalstege mit Nuppen in den Zwischenräumen, nach oben hin abgesetzt durch einen Ring feiner Punktnuppen. Graubräunliches Glas. Die Schlangen aus fein geripptem farblosen Glas, bernsteinfarbene Augen, Kämme und Ansätze grün.

Berlin, Kunstgewerbemuseum Schloss Köpenick.
Inv. Nr. 09,249

III 10g
Abb. 52b

Pokal

Zuschreibung und Datierung wie III 10a, b;
H. 22,1 cm

Allgemeiner Aufbau, und Kuppaform wie III 10c.
Fadeneinlagen weiss-rot-gelb, Ansätze grün.

Stockholm, Nationalmuseum. Inv. Nr. NM 19/1902
Schenkung W. Bendix 1902 (seit 1887 im Museum deponiert)

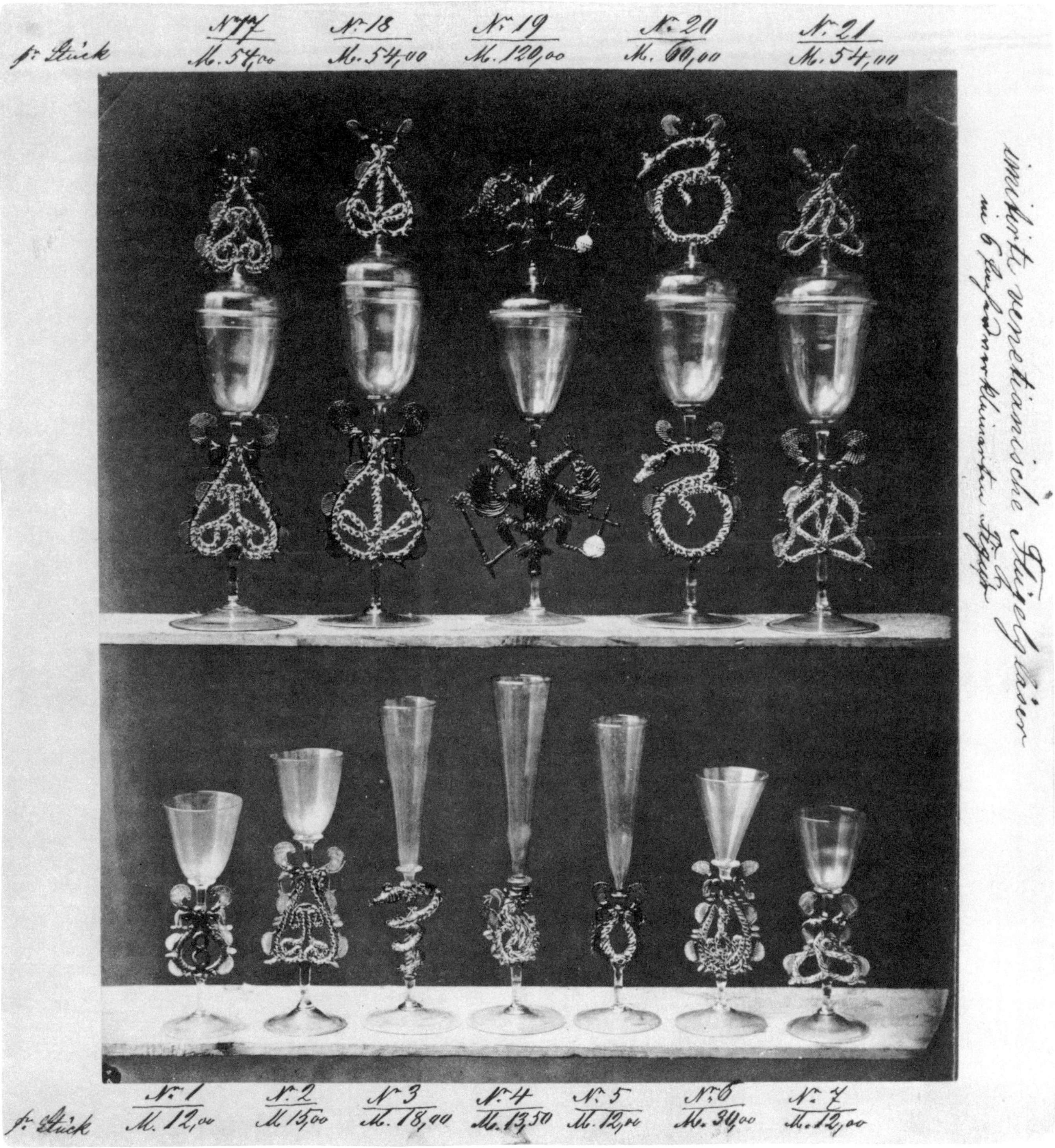

ABB. 53. *Musterfoto der Werkstatt C. H. F. Müller, Hamburg, 70er Jahre 19. Jh.*

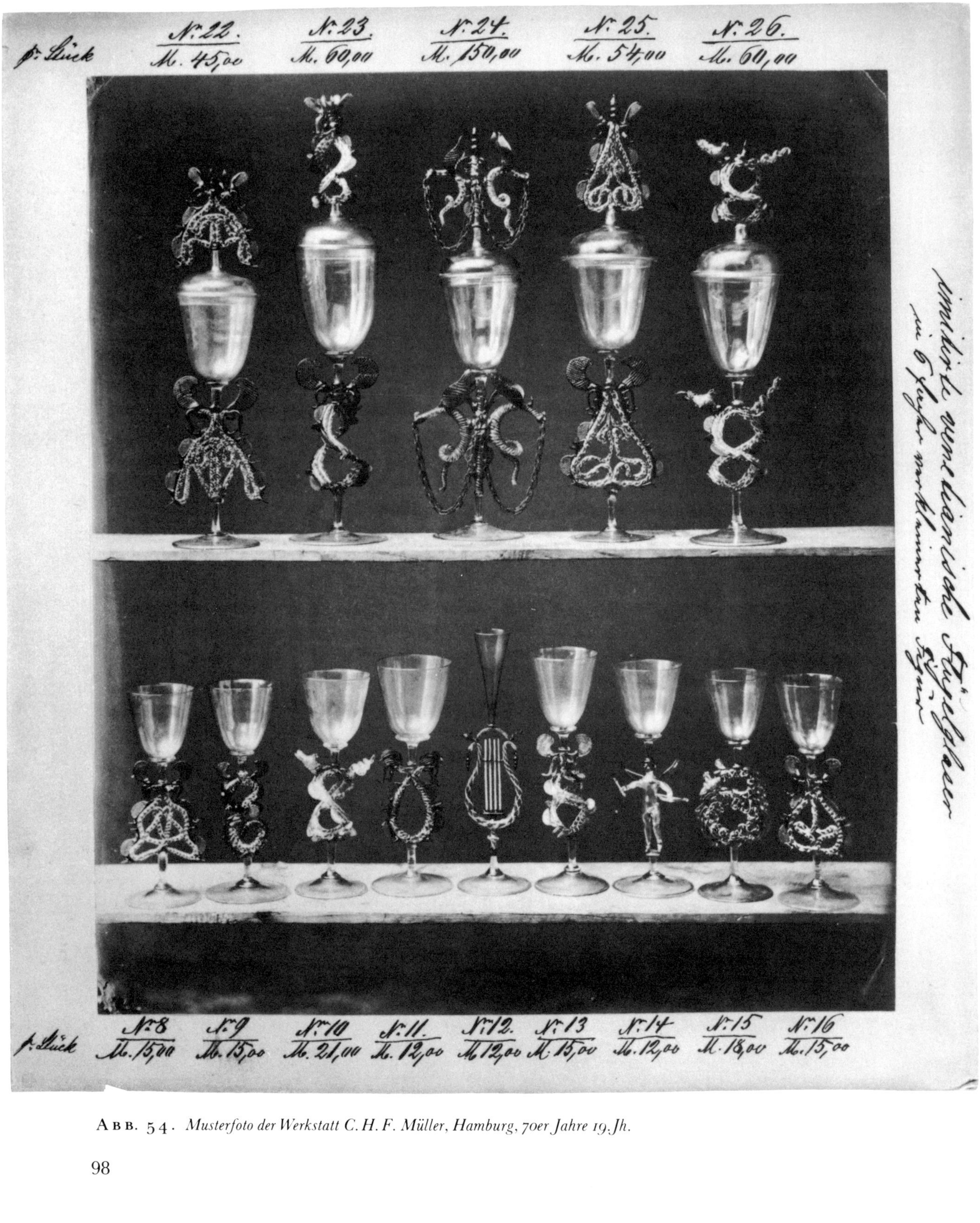

ABB. 54. *Musterfoto der Werkstatt C. H. F. Müller, Hamburg, 70er Jahre 19. Jh.*

Abbildungsnachweis

Nr. 1, 32, 34, 35, 40, 42, 53, 54 Museum für Kunst und Gewerbe, Hamburg. — Nr. 2, 16 Victoria & Albert Museum, London, Crown Copyright. — Nr. 5, 9, 13, 52 a Foto-Commissie Rijksmuseum Amsterdam. — Nr. 8 Kulturen, Lund. — Nr. 10 British Museum, London. — Nr. 11 Metropolitan Museum, New York. — Nr. 12 Musée Jacquemart-André, Paris. — Nr. 18, 19 Gewerbemuseum der Landesgewerbeanstalt Bayern, Nürnberg. — Nr. 21 Österreichisches Museum für angewandte Kunst, Wien. — Nr. 22 Walter Danz, Halle. — Nr. 24 Peill & Putzler Glashüttenwerke, Düren. — Nr. 26, 34/2, 43, 47 The Corning Museum of Glass, Corning, N.Y. — Nr. 27, 33, 37, 45, 49, 51, 52b Nationalmuseum Stockholm. — Nr. 28 Zeichnung Architekt Fritz Ricke, Hannover. — Nr. 32/2 Gerd Remmer, Flensburg. — Nr. 41a Teigens Fotoatelier, Oslo. — Nr. 41b Hermann Friedrich, Wülferode-Hannover. — Nr. 48 Landesmuseum für Kunst- und Kulturgeschichte, Vocke Museum, Bremen. — Nr. 50 Museum für Kunsthandwerk, Frankfurt/M. — Nr. 52c Kunstgewerbemuseum, Berlin Köpenick. — Die übrigen Abbildungen nach Aufnahmen des Verfassers.

Dank

Meine Bemühungen um eine erste Zusammenstellung der Werke C. H. F. Müllers wären ohne die Hilfe einer grossen Zahl von Kollegen unvollständig geblieben. Für Hinweise auf mir unbekannte Gläser möchte ich insbesondere danken: Herrn Dwight Lanmon, Corning, Frau Inger-Marie Lie, Oslo, Frau Jessy McNab-Dennis, New York, Herrn P.C. Ritsema-van Eck, Amsterdam, Frau Barbara K. Schnitzler, Cambridge, Herrn Rainer Schoch, Darmstadt, Frau Anna-Elisabeth Theuerkauff-Liederwald, Berlin und Herrn Dag Widman, Stockholm.

Für freundliche Hilfe bei der Beschaffung von Fotografien und für Detailinformationen habe ich zu danken: Herrn Jörn Bahns, Nürnberg, Frau Margrit Bauer, Frankfurt, Herrn A. L. den Blaauwen, Amsterdam, Frau Nicole Blondel, Paris, Frau Elisabeth Bornfleth, Nürnberg, Herrn Robert Brill, Corning, Herrn Robert Charleston, London, Frau Olga Drahotova, Prag, Herrn Astone Gasparetto, Venedig, Herrn Fritz Goldkuhle, Bonn, Frau Gisela Haase, Dresden, Herrn Heinz-Wilhelm Haase, Bremen, Herrn Wolfgang Hennig, Berlin-Köpenick, Frau Helga Hilschenz, Hannover, Herrn Hermann Jedding, Hamburg, Frau Christel Mosel, Hannover, Herrn J. V. G. Mallet, London, Frau Riet Neerinx, Arnhem, Frau Waltraut Neuwirth, Wien, Herrn Nils Nilsen, Lund, den Herren Opeldus und Eilken, Firma C. H. F. Müller, Hamburg, Frau Maria Luisa Righini Bonelli, Florenz, Herrn Axel von Saldern, Hamburg, Herrn Helmut Scherf, Eisenach, Herrn Ulrich Schulte-Wülwer, Flensburg, Frau Gisela Schulz, Hamburg, Herrn Hugh Tait, London und Frau Wibeke Woldbye, Kopenhagen.

Besonderen Dank schulde ich der interessierten und selbstlosen Unterstützung durch Herrn Kurt Wallstab, Glasgestalter in Darmstadt/Griesheim, dessen Hilfe bei der Klärung technischer Fragen unschätzbar war, nicht zuletzt durch die praktische Demonstration komplizierter Blasvorgänge.

THE USE OF EQUILIBRATED SILICA GEL FOR THE PROTECTION OF GLASS WITH INCIPIENT CRIZZLING

ROBERT H. BRILL

IN A PREVIOUS PUBLICATION[1] we have discussed glasses showing incipient crizzling, that is, glasses in the earliest stages of crizzling. This condition threatens many pieces of European, East Asiatic, and American glasses manufactured between the seventeenth and nineteenth centuries. Examples exist in public and private collections throughout the world. These glasses have absorbed moisture during their lifetimes, becoming hydrated to a depth of approximately 50 microns (0.05 mm). For some reason, however, possibly because they have been stored more-or-less continuously in moist environments, they have not developed the same severe surface crazing as that which characterizes the more familiar condition described simply as "crizzling." On glasses showing incipient crizzling, the damage may often be overlooked, for only when the objects are viewed in just the right light do the silvery, crazed patterns of early crizzling become evident. Nonetheless, these glasses are very unstable, and if exposed to low relative humidities for extended periods of time (ranging, perhaps, from several months to a few years) the glasses slowly release their water of hydration. This drying-out is accompanied by shrinkage in the affected layer which leads to the opening of fissures within that layer. The glass is thus transformed into a fully-crizzled state marked by a loss of transparency. If this degenerative process proceeds far enough, not only do the objects become unsightly, but a loss of surface material may also occur. The process is not unlike that which produces crackling when a layer of mud dries out in strong sunlight.

The chemical effects causing crizzling were discussed at length in the earlier publication, along with experiments in which the effects were reproduced in the laboratory. It was verified that the primary cause of crizzling is a deficiency in lime in the glass composition or an excess of alkali, two conditions which accelerate the hydration of glasses. Because it is believed that the degeneration of a glass with incipient crizzling into a severely crizzled state is the result of the loss of water of hydration, it seems reasonable to assume that in order to prevent degeneration, glasses with incipient crizzling should be stored or exhibited in relatively moist conditions, so as to inhibit loss of their water. With that aim in mind, storage at

1. Robert H. Brill, "Crizzling—A Problem in Glass Conservation," *Conservation in Archaeology and the Applied Arts*, London: The International Institute for Conservation of Historic and Artistic Works, 1975, pp. 112–134. The first stage of this research, that described in the above publication, was conducted in part under a grant from The National Endowment for the Arts.

constant and moderate levels of temperature and RH was recommended. Moreover, it was recommended that dramatic spotlighting and regular cycling of temperature (as might be brought about by turning on and off gallery lights according to a daily schedule) both be avoided. It was concluded that ambient temperatures comfortable to people were fully satisfactory and that an RH of 45–47% should be the target, although any RH within a range of 40–55% would probably be quite safe. Short-term excursions to about 35% or 65% were thought to cause little harm, although long-term storage outside those limits might be harmful. From what has been learned in the meantime, these guidelines still seem valid.

It must be emphasized that by storing glasses under these conditions one can only hope to forestall the degeneration of incipient crizzling into severe crizzling. Storage under these, or under any other conditions which we can presently imagine, is not going to undo the damage already inflicted upon the glass, nor is it going to correct the inherent chemical instability and susceptibility to future degeneration.

It is the author's opinion that the same target conditions of temperature and RH are probably also as safe as any other conditions for the storage or exhibition of already-crizzled glasses. We see no advantage, and indeed see distinct disadvantages, to unnecessarily subjecting already-crizzled glasses to markedly more moist or markedly drier conditions than those specified for glasses showing incipient crizzling.

After the requirements for safe storage had been established, a schematic plan for an exhibition case intended to attain these conditions was prepared. The second stage of our work was to construct such a case and monitor its temperature and RH through a full year's cycle. That stage has now been completed, and the results are presented here.

The next stage of our work will be to design cases for glass installations in the new galleries of The Corning Museum of Glass which will open in 1980. It is hoped that the galleries themselves will be maintained between limits of about 35 and 65% RH as part of the architectural specifications, but there is no assurance that these will be met, and there are likely to be widely ranging short-term fluctuations. These conditions should probably suffice for the major parts of the glass collection which are not very susceptible to damage by RH changes. However, the crizzled glasses (especially those with incipient crizzling), certain glasses with painted or gilt decoration, and glasses with particularly fragile weathering crusts will require several individually controlled cases. Our present intention is that these cases will be controlled by equilibrated silica gel according to the guidelines listed below.

SUMMARY OF FINDINGS

The discussions which follow are likely to be of interest mostly to conservators and museum scientists. Therefore, so as to aid other readers, we present a summary of our findings here at the beginning of this article, along with some recommendations.

Although a great deal of useful information has been gathered, this stage of the project has been only partially successful.[2] We have demonstrated, to our own satisfaction, that storage in a case controlled by equilibrated silica gel is a valid, safe, and practical means for preventing glass with incipient crizzling from degenerating into a severely crizzled state. All the advantages anticipated for the system seem to have been realized. However, the experimental case used in the most important experiment, while it proved effective in isolating

2. We have certain reservations about publishing the results of only partially successful work, but the system seems as promising as ever, providing that proper corrections are applied, and the results complement nicely the findings of Thomson. Moreover, it will be another year before we can gather additional useful data, and we feel that others might benefit from knowing our experiences now.

the glass from high exterior ambient humidities prevailing during the summer months, showed a persistent drift downward in relative humidity during the dry winter months. Because the dry winter months are the time when protection is most needed, questions arose about the usefulness of the system.

The main point of concern was whether the drift was the result of simple leakage—as was expected—or whether there was some fundamental flaw in the principle of the system. The appearance of two fine articles related to the use of conditioned (equilibrated) silica gel systems as this manuscript was in its late stages of preparation have pretty well settled our doubts. These two articles are by Nathan Stolow of the National Museum of Canada in Ottawa[3] and Garry Thomson of the National Gallery in London.[4] Both emphasize the problems of case leakage and provide guidelines for the quantities of gel required to compensate for leakage. It is apparent to us now not only that our case was not well enough sealed, but that the reservoir of silica gel was too small to prevent the downward drift under the extremely dry conditions prevailing in our galleries.[5]

Our findings and recommendations (stated in the form of intentions for our own new installations) are summarized as follows:

1. The equilibrated silica gel system is a valid, safe, and practical means for controlling RH in cases used for the storage of glass and is recommended for preventing the degeneration of incipient crizzling into full crizzling.

2. Leakage is a major problem. In all but the most extraordinary conservation systems, where the expenditure of extraordinary amounts of money and effort are warranted, it is wise to consider leakage inevitable. Instead of attempting to stop leakage altogether, it is probably much more practical to minimize it and compensate for the leakage that does occur. (One difficulty is the lack of any straightforward way of measuring leakage and therein lies part of the value of Thomson's theoretical treatment, which permits one to come to grips with the leakage problem.)

3. In our case designs for the Museum's new building, care must be taken to provide a sufficiently large quantity of equilibrating gel following the guidelines of Stolow and Thomson. The quantity of gel should be of the order of 20 kg per cubic meter of total case volume. (This is a four- to five-fold increase over what was used in our experimental case.)

4. It is probably more practical to use a few small exhibition cases than one large case. We will aim to have no controlled case larger than about 1.0–1.5 cubic meters in volume (including the equilibrating chamber).

5. The silica gel will be spread so as to present a large effective surface area to the air within the cases, and will be located, as much as possible, in close proximity to those objects it is intended to protect. Depending upon the nature of the individual cases, the gel will either be incorporated into fabric-covered panels or will be placed in open trays partially hidden by perforated surfaces. While we would like to minimize the presence of water-absorbent materials other than silica gel, on the basis that they introduce other variables into the system, those materials which are necessary will be selected with a view toward providing further beneficial buffering of the RH.

6. A separate conditioning apparatus will be constructed for the equilibration and reconditioning of the silica gel, and a back-up reserve of conditioned gel will be maintained.

7. A remote sensing RH and temperature probe system, similar to that described here,

---

3. Nathan Stolow, "The Microclimate: A Localized Solution," *Museum News*, 56, No. 2, November/December, 1977, pp. 52–63. This publication also serves as an excellent entry into the bibliography on this general subject.

4. Garry Thomson, "Stabilization of RH in Exhibition Cases: Hygrometric Half-Time," *Studies in Conservation*, 22, No. 3, August, 1977, pp. 85–102.

5. As these words are being written, during the Great Blizzard of February 1978, the RH in our Museum's main gallery is about 12%.

will be used to monitor case conditions and ambient conditions.

8. All lighting devices will be external and well-vented, with heat shields installed. No spotlighting will be used for dramatic effects unless it can be demonstrated that it produces no temperature increase in the objects.

9. We will consider the use of a "double-walled" barrier to attain longer hygrometric half-times. Thomson has shown that the hygrometric half-time depends upon the difference between the internal and average external RH. If an enclosed exhibition space is surrounded by another enclosed space containing silica gel equilibrated to the same target RH, then leakage from the inner chamber actually holding the glass objects should be much slower, because the gradient between that chamber and the surrounding chamber will be much smaller.

A few further remarks on the use of silica gel are worth repeating. Most readers already will be familiar with the use of silica gel as a desiccant, that is, as a material intended to absorb all the moisture from its surrounding air and leave the air as dry as possible. If the silica gel is itself dry to begin with, this is exactly how it behaves, because it has such a strong tendency to absorb water. In the present application, however, the silica gel used has already had water added to it in proportions of the order of 20–30 percent by weight. Hence, its tendency to absorb water has been much reduced. Silica gel containing this much water does not produce a near-zero RH in the surrounding air but instead, establishes an equilibrium RH of about 40–50 percent in the surrounding air. If the gel is the most abundant phase in a closed system and has a large surface area exposed, it will either remove moisture from the air or release moisture into the air until its own equilibrium RH has been established. Silica gel is well-suited for controlling exhibition cases because it is easy to handle, can be regenerated, does not become messy to handle in the hydrated state, and is chemically inert. In addition, its equilibrium RH, for a given water content, is said to be relatively insensitive to temperature changes.[6]

As with all conservation treatments or procedures, however, the silica gel system must be employed judiciously. In this instance, *judiciously* takes on a very important meaning because there is one way that the system could go seriously wrong.

As was described above, the whole efficacy of this system centers around one particular property of silica gel, namely, that it will release or take up moisture to maintain some constant RH in the surrounding air. This property of the gel is characterized quantitatively by its equilibrium Rh *vs.* moisture content curve, which describes what surrounding RH corresponds to what water content of the gel. During the course of this work, we have come to appreciate that this curve is not necessarily the same for all lots of silica gel. In fact, buying silica gel might be much like buying vegetables. A basket of corn bought in Petunia Pump last year might be quite different from a basket of corn bought in Pathan next spring, even though both are quite properly called "corn." So, for the present at least, we feel each new lot of silica gel should be regarded as a unique entity, and unless the manufacturer has provided an EMC *vs.* RH curve (an adsorption isotherm) the user must determine one himself.

Our recommendation to those who might consider making use of silica gel is, by all means, to go ahead but to be certain that the installation is supervised by a conservator or someone else who has a general understanding of how silica gel behaves. It is important first that care be taken to prepare the gel properly and to be certain that it is equilibrated to the target RH desired. (This process might take as long as a few weeks.) Then it is crucial that

6. Nathan Stolow, "Fundamental Case Design for Humidity Sensitive Museum Collections," *Museum News*, 44, No. 11, February, 1966, pp. 45–52.

the RH of the system be reliably monitored. The danger is that if silica gel which is not sufficiently moist is used, or if the gel is allowed to dry out somehow, it could have the effect of drying out the glass and thereby cause harm. But this is not reason for alarm, for as long as the RH of the system is monitored, that should never happen. If the RH shows signs of drifting downward, the drift can be very simply corrected by adding water to the system; if it drifts upward, drier silica gel can be added. (In actual practice, this usually would be done by adding freshly-equilibrated silica gel.)

DESCRIPTION OF AN EXPERIMENTAL CASE

Because of budgetary limitations, we did not construct new cases for our experiments, but instead modified existing cases. At the outset it was realized that the results would not be ideal, and that the cases would not meet the optimum specifications. Nevertheless, the experiments served well to evaluate realistically a number of variables.

Our most important experiment made use of an aluminum-framed case with permanently-fixed glass walls on three sides and four large sliding glass doors on the front. The original top and bottom of the case were also of plate glass. The inside dimensions of the case are: height 1.66 m, width 2.87 m, and depth 40 cm. The volume of the exhibition case proper is approximately 1,900 l. It rests upon a steel base supported by six steel legs with a clearance of 45 cm beneath. The case contains three plate glass shelves mounted on adjustable steel racks.

The main advantage of this case for our experiment was that because it is made of plate glass and metal, it did not present surfaces which were highly absorbent to water vapor. The presence of fabrics or wood in a case is a complicating factor, as is demonstrated below in an experiment with a different case. The main disadvantage of the case was its sliding glass doors which were notoriously loose-fitting. The doors were sealed during the experiment with lengths of plastic tubing, used as gaskets, and duct tape. While this afforded a seal of sorts, it proved inconvenient to remove and replace the tape on those occasions when the case had to be opened. The interior seams of the case were caulked with a silicone caulking which worked fine, because it remained pliable and, therefore, would not have cracked if the case had shifted. (No build-up of fumes released from the caulking was detected.) Despite the effort taken, however, we know now that the case was not sealed nearly as well as we would have liked.

The original glass bottom was removed for the experiment, and a tightly-fitting, brazed sheet-metal compartment was fastened permanently in place beneath the case to serve as the chamber for the silica gel. Access to this lower chamber is through two gasketed, metal faceplates, held in place with sheet-metal screws. The exterior of the metal chamber was painted, but the interior surfaces were not.

The lower chamber contains six sheet-metal racks, each of which supports four enamel trays measuring 39.0 cm by 17.5 cm by 5.0 cm. The trays hold the equilibrated silica gel. In the experiment described, the silica gel filled the trays to a depth of approximately 1.5 cm. The exposed surface area, not corrected for granularity, was about 1.2 m^2. The trays were easily removed whenever necessary by sliding the racks out through the opened faceplates. The lower chamber is separated from the case itself by a polyethylene grid which provides adequate support for objects to be placed upon it while also permitting a free exchange of air between the lower chamber and the case proper. The volume of the equilibrating chamber is approximately 400 l, so that the total volume of the system (exhibition chamber plus equilibrating chamber) is approximately 2,300 l.

During our experiments, it was found that no fan or circulating device was necessary, as there seemed to be sufficiently free exchange of air between the two compartments. This was helpful because it meant there was one less thing to go wrong and one less item in the system requiring maintenance. This case was not lit except by ambient room lighting, so no precautions were required against heat buildup from that source.

THE SILICA GEL

The silica gel used for our experiments had been prepared earlier by Dr. David Fischer, who worked at the Museum from January, 1973 to December, 1975 with the assistance of Mr. Joseph Maio, of the Museum's staff. In all, they equilibrated about 145 kg of silica gel. When purchased, the gel was nearly dry. Deionized, distilled water was sprayed directly onto the gel to hasten its absorption because storage with liquid water in a closed compartment proved to be too slow a means for equilibrating the gel. (Caution must be exercised when adding water to silica gel because heat is evolved by the process.) The amount of water added was calculated from data published by Stolow.[7] The curve reported in Dr. Stolow's publication gives relative humidity values for various equilibrium moisture contents of silica gel. (The equilibrium moisture content, or EMC, is the water content expressed as a percentage of the weight of the *dry* gel.) In Dr. Fischer's preparations, adjustments had to be made, which we assume resulted from differences between the silica gel we were using and that used by Dr. Stolow. The separate batches of moistened gel were then stored in double-thickness plastic bags inside galvanized metal cans which were then sealed with tape. Dr. Fischer measured equilibrium RH values for samples of the separate batches after the air above them in closed chambers had come to a constant RH. The results ranged from 43.5% to 47.8% RH. Measurements made two years later on the batch of gel actually used in our case experiment[8] agreed to within ±0.5% RH with Dr. Fischer's measurements.

Silica gel of 12–28 mesh was used. This was a compromise between two effects. The smaller the gel particles, the greater the exposed surface area and, consequently, the more rapid the equilibrium time. But it was thought that gel of too small a particle size might produce dust problems throughout the case. Actually, although we carried out tests specifically to detect dusting, none was found, and it was concluded that for the way we had introduced the gel (in horizontal trays), the 12–28 mesh size was fine. However, Dr. Stolow recommends using a larger size, 3–8 mesh, as mentioned below. (See A Final Note.)

A curve for equilibrium RH values *vs.* EMC for the gel used in this experiment was estimated by adding known weights of water to known weights of the gel, which had been dried previously at 190°C for seven hours. The water was added to the gel by evaporation transfer inside sealed 9.7-liter desiccator chambers. After the water had completely evaporated and the RH in the chambers had become constant—a process which required several weeks in some instances—the systems were assumed to have come to equilibrium. Samples of the gel were removed, and their water contents were determined by drying to constant weight at 180°C. The resulting EMC *vs.* RH curve is shown in Figure 1. Also plotted on this graph are data for three samples of silica gel removed from the Crizzling Case during an experiment described below.

The curve is displaced somewhat from the similar curve published by Stolow. Because we do not have a sufficient number of points for plotting both ascending and descending

7. Stolow, *loc. cit.*, p. 51. See also, A Final Note.

8. Davison, Code No. 408-08-72-201, Grade 408, Mesh 12-28, Package 9445.

branches reliably, we have elected not to indicate hysteresis. Some of the points on the curve drawn were probably crossing over from the ascending branch toward the descending branch, and the three points for the samples withdrawn from the case should be on the descending branch.

The rather sharp bend in the curve near EMC -25% is of special interest. One wonders if such a plateau might result from saturation with water of hydration corresponding, for example, to $H_2SiO_3 \cdot H_2O$ ("EMC" -23%), $H_2Si_2O_5 \cdot 2H_2O$ ("EMC" -26%), or some other stoichiometric composition. Unfortunately, because of a very tight publication deadline, we have not yet been able to verify these data by repeat runs—and we confess to not having made a thorough literature search on the subject.

TEMPERATURE AND RELATIVE HUMIDITY MONITORING

The temperature and relative humidity of the sealed experimental case were measured by two sensing probes placed inside the case at diagonally-opposed corners. The probes were attached by cables to a humidity-temperature reader manufactured by the Phys-Chemical Research Corporation.[9] Although the system is designed to incorporate only one sensor probe, a special switch box was made which allows the system to accommodate twelve different sensors. These are attached by cables of various lengths (measuring up to 28 m) spread out in a network to various locations in the Museum where we chose to monitor the temperature and RH. The sensors were all individually calibrated by the manufacturer, and those used in the experiments described here were checked periodically against a sling psychrometer. The two probes in the experimental case check one another to within 2% of RH units.

It was decided not to record the data automatically, partly because it would have been very expensive, and partly because it would have been difficult to do that simultaneously for several sensors. In any event, there was no need for recording data automatically. For routine monitoring, four readings of temperature and RH were taken between 9 a.m. and 5 p.m. daily on each of the probes in the experimental case. During the hours when the Museum offices were closed, and on weekends, readings were taken by the Museum security guards on their rounds. Normally, four readings were taken each night. As each reading was taken, a notation was made of the weather conditions. When experiments had just been started or at particularly important stages, more frequent readings were taken. It is such an easy matter to collect data—one simply presses a button and reads the temperature and RH off two dials—that it is quite convenient to obtain nearly continuous data, providing only that one is willing to stay near the read-out meter for the interval required. The temperature and RH of the ambient air immediately outside the test case were recorded continuously on a recording hygrothermograph[10] which was in reasonably close calibration with the sensors, but which did not yield data of quite as high precision and accuracy.

In retrospect, the number of data taken during the course of the research described below—the Crizzling Case experiment, the monitoring of the Ravenscroft Case, the monitoring of three stations in the galleries, and the determination of the EMC *vs.* RH curve—is quite astounding. Some 20,000 individual readings of RH were recorded, along with an equal number of temperature measurements. This attests to the facility with which the readings

9. The instrument is a Humitemp-2 equipped with PCRC-11 HPT sensing probes. The address of the manufacturer is Phys-Chemical Research Corporation, 36 West 20th Street, New York, New York, 10011.

10. The instrument is a familiar one: model no. 5-594-2 manufactured by the Belfort Instrument Corporation of Baltimore, Maryland.

are taken, for it never seemed at all a bother, and also emphasizes our indebtedness to the security guards who collected half of these data.

THE "CRIZZLING CASE" EXPERIMENT

The case described above was closed on August 25, 1976, after 12 kg of equilibrated silica gel had been spread out in the enameled trays. It was monitored according to the schedule outlined above (about nine readings per day) until September 20, 1976, when the relative humidity had leveled off at 44%. Then, 120 objects of glass, 15 with incipient crizzling and 105 already-crizzled pieces, were placed in the case, and it was resealed. The presence of the objects in the case appeared to cause the RH of the enclosed air to drift downward over the course of three weeks to a level of 40%. Evidently, the crizzled glass absorbed moisture from the air more quickly than the silica gel could replenish it. At that point, liquid water was added to the system in order to elevate the RH. This had to be done five more times over the succeeding year to keep the RH from dropping below the 40% level. In most instances, the water was sprayed directly onto the surface of the gel (see A Final Note, below), but occasionally it was added as liquid water in small beakers distributed throughout the case. This latter procedure is definitely not recommended (see A Special Precaution, below). Whenever water was added, the gel was always lightly raked to expose a fresh surface. Sometimes a slight crustiness was found to have developed on the surface of the gel after it had been in the case for several weeks. This was accompanied by a faint yellowish coloration.

From the above it is clear that the presence of this large number of crizzled objects with their large effective surface area altered the RH within the case, and that this effect is not negligible. This is to be expected, because where two absorbent phases are present within a closed system, the actual RH established should be intermediate between the individual equilibrium values for the two phases. If the equilibrium RH values of the two phases are very different, and if a much greater effective surface area of either phase is present, the net result would be that water would be "pumped" from one phase to the other, forcing the less abundant phase to absorb or desorb moisture until it could tolerate the equilibrium RH demanded by the more abundant phase. In this experiment the equilibrium RH values of the crizzled glass[11] and of the silica gel are quite close to begin with, and the effective surface areas seem well balanced against one another, so no extensive transfer of water from one phase to the other occurred.

The case was taken through more than a full year's cycle during which it became well stabilized for nearly six months from April until October at a level of 40–43% RH. During that interval, the ambient RH changed from 40% to 65% back to 40%. During the winter months, however, a gradual drift downward could be seen, and it was this which had to be compensated by the addition of liquid water to the system. From mid-October until mid-March, when the central heating was on, the RH just outside the case was rarely above 25% and frequently below 20%, sometimes remaining for more than a week at a time at 12–14%. We attribute the drying out of the case to exchange with this very dry air by leakage. The temperature and RH history of the experimental case is summarized and plotted in Figures, 2, 3, and 4.

Perhaps the most significant point is that the silica gel buffered the RH changes in precisely the way it was intended to do. Setting aside for the moment the drift caused by leakage, it is very satisfying to see that the short-term fluctuations are so small. It will be recalled that one of the advantages of using silica gel is

11. See Miscellaneous Notes.

that its equilibrium RH values are said to have but a small temperature dependence, so that when the temperature changes, the gel still tries to maintain virtually the same RH. This was borne out in several instances where after abrupt temperature changes occurred, the RH was restored to its starting value, lagging behind by only a few tens of minutes.

During the course of these experiments, no visible changes occurred in the objects in the cases—but then none were anticipated. (See the exception in the note below.) The changes which would occur in crizzled glasses or glasses with incipient crizzling over a year would not be expected to be noticeable at the RH level maintained. That, after all, was the point of the experiment. However, changes on a microscopic scale should be monitored in future experiments. Some such experiments were underway at The Corning Museum of Glass in 1971, wherein tapes were applied to make microscopic replications of crizzled surfaces, and the tapes were recorded by electron microscopy. However, the original tapes were destroyed in the 1972 flood, and the work has not been revived, because there is a certain danger in such experiments. To put a tape on the surface of a crizzled glass is to risk the removal of any loosely-attached material on the surface, and we have been hesitant to run that risk.

THE "RAVENSCROFT CASE"

As part of a special exhibition mounted at the Museum in 1976, a fine example of English glass, a goblet made by George Ravenscroft in 1676, was exhibited (CMG no. 50.2.2). This vessel bears a raven's head seal, which Ravenscroft used to mark his later glasses made with a corrected formula. These glasses, he supposed, would not be susceptible to crizzling, as had some of his earlier glasses.[12] Nevertheless, Ravenscroft's corrected formula did not produce really stable glasses, and this particular object shows clear signs of incipient crizzling. It is a prime example of the sort of glass which must be carefully protected. If it is not, it will surely deteriorate into a severely-crizzled condition.

The object was to be exhibited in a small case constructed of plywood. The exhibition compartment volume was 68 l. The case was designed originally to be lit from above with spotlights, since it was to be located in a darkened gallery. The interior of the case was lined with about one square meter of purple felt, which complicated control of the relative humidity, as the felt tended to absorb a great deal of moisture. The case was made reasonably air-tight by sealing it with duct tape around all seams. A sheet-metal trough covered with the same felt and lined with polyethylene bags was placed in the rear of the exhibition compartment. Then 1,500 g of equilibrated silica gel was placed in it. A sensor was suspended unobtrusively in the rear of the case, and the case was carefully tested before the Ravenscroft goblet was installed. A mock-up of a colorless glass laboratory beaker was placed where the goblet was to be and lit with a twelve-volt lamp of a sort which frequently has been used throughout the Museum. The temperature readings as measured by a thermistor thermometer—which accurately measures the temperature of the glass, and not just the surrounding air—showed that illumination with this light raised the temperature of even this colorless glass from 27.0 to 37.5°C. This was judged to be completely unacceptable for illumination of the Ravenscroft glass in its unstable condition. When a plate of heat-absorbing glass and a piece of diffusing material were placed between the lamp and the exhibition chamber proper, the surface temperature of the glass dropped to 32.5°C, which was still

12. There are numerous accounts of Ravenscroft's early experimentation. One is: A. Hartshorne, *Antique Drinking Vessels*, New York: Brussel and Brussel, 1968, pp. 240–241.

regarded as a bit high, and which was still warmer than the air in the chamber. The spotlight was replaced with a fluorescent lamp having external circuitry and the temperature then held at about 25°C, and did not differ measurably from the temperature of the surrounding air. Therefore, the modified lighting was deemed safe for the Ravenscroft object.

In an effort to make the exhibit more instructive, a volume from the rare book collection of the Museum Library was placed in the case with the goblet. This is a 1662 edition of Neri's *The Art of Glass* (Special Collection no. 327). The volume was opened to a page relevant to the chemistry of the Ravenscroft goblet. When placed in the case, the book produced a startling drop in RH. It absorbed water vapor rapidly, because it had been shelved under low-humidity conditions for several months prior to being incorporated in the exhibition. This is dramatic evidence that the presence of additional absorbent materials—in this case the felt and the book—has an important effect on the RH finally attained. Even as effective a controlling agent as the silica gel had a hard time coping with these complications. If one is attempting to control a case with a material like silica gel, we believe it is inadvisable to introduce any strongly absorbent materials unnecessarily. For our installation of the Ravenscroft object, water was added to the system in order to elevate the RH somewhat. After the addition of 65 g of water to the 1,500 g of gel, a safe and stable RH of 42–45% was established and maintained for twenty-four weeks, after which the exhibition was dismantled. The Ravenscroft object appears to have been completely unaffected by its exhibition experience.

THE HYGROMETRIC HALF-TIME

Garry Thomson's treatment of the stabilization of relative humidity in exhibition cases[13] makes use of the concept of a hygrometric half-time, that is, the time required for the RH of a case to drop halfway to the RH level of drier ambient air. His mathematical treatment assumes an exponential decay of the RH in the case and takes into account the rate of air leakage and characteristics of buffering materials within the case. We have tested Thomson's model by comparing it to the empirical data collected during the monitoring of our own Crizzling Case. Thomson's model predicts that a plot of ln(C-R) *vs.* t should yield a straight line. (C is the RH within the case at time t and R is the RH of the surrounding air.)

A plot of our data for the interval between December 12, 1977 until January 23, 1978 (Fig. 5) does indeed yield a straight line verifying that the decay is exponential. (R is known to have averaged about 22% during this period.) From the slope of this plot the decay constant, k, was found to be 0.0192 per day, and the calculated half-time is 36.1 days. From our EMC *vs.* RH curve an average value of M -5.9 was assumed, and in our experiment, B is known to have been -3.5 kg/m³. Therefore, by using Thomson's equation *vi*[14] and our own experimentally determined rate constant (k), it was possible to calculate that during that interval our case was undergoing about 2.3 air exchanges per day with the ambient air. This figure seems somewhat high and may not be real, because there could have been another factor contributing to the downward drift. Water had recently been added to the system and the silica gel might still have been in the process of reequilibrating. This would cause a downward drift if the rate of diffusion of water into the interior of the gel grains was relatively slow.

In our designs for cases for the new museum building, we will make use of Thomson's model

13. Thomson, *op. cit.*

14. Thomson, *ibid.*, p. 96.

$k_N = \frac{0.173}{MB + 0.173}$ where M = moisture gain in g per kg for 1% RH rise and B = dry weight of buffer in kg/m³.

and equation by increasing B, the quantity of buffering gel used, and attempt to reduce k by improving sealing so as to reduce leakage. In addition we shall explore still another approach.

Thomson's explanations make clear the fact that the leakage rate of the case is dependent upon the RH gradient between the case and its surroundings. His treatment quantifies the intuitive notion that the driving force behind leakage is the difference in RH and suggests that one way of extending the half-time of the drying process is to find some means to reduce the term (C-R). This, in turn, suggests that the presence of an air envelope of RH close to C surrounding the case would inhibit loss of moisture from the case and brings to mind the possibility of using a double barrier to further stabilize the case. This is just another way of looking at the rather obvious fact that a double-walled container would be slower to dry out but emphasizes that this would be especially true if the space in between also contained buffering material.

We have explored this possibility from a mathematical viewpoint and believe that a considerable extension of hygrometric half-time could be realized. The ultimate value of such a system would be determined by balancing the advantage of increasing the half-time against the inconvenience and expense of constructing a more elaborate case. Such a balance can be made favorable only by the ingenuity of the case designer.

SOME MISCELLANEOUS NOTES

"The Vapor Pressure" of a Crizzled Glass

A group of five badly-crizzled eighteenth-century French goblets was placed in a chamber with one of the same sensor probes used for monitoring the crizzling case. The chamber used was a 9.7-liter laboratory desiccator, although only the crizzled glass was placed in the chamber and no desiccant or other materials. When the experiment was started the relative humidity was 46%. The RH slowly drifted downward and after eight weeks, leveled off at 41%. If one disregards the effect of absorption and desorption of water vapor on the walls of the chamber itself, it might be inferred that the "vapor pressure" of the adsorbed water and water of hydration of this particular crizzled object is ~7.2 mm Hg, or an RH of 41% at 20°c. This corresponds, within experimental error, to the value of 42% given as the equilibrium RH for the monohydrate of potassium carbonate cited by Robert Organ.[15] It also makes sense that the vapor pressure of water on—and/or in—a crizzled glass should be of the order of 40% or greater. Historically, the glass would have become hydrated to the extent that its equilibrium vapor pressure, expressed as an equilibrium RH, would be close to that of the average ambient RH it experienced over the past three centuries, and that value probably would have been of the order of 45%.

In a separate experiment, one of these goblets was heated at 110°c for one hour and its loss in weight, presumably consisting of evolved adsorbed water and water of hydration, amounted to 0.13 g or 0.12% of its weight.[16] The surface area of the goblet is about 278 cm^2, so its water content had been about 0.00047 g/cm^2. The same object was then saturated with water by letting it stand in a chamber at 100% RH for several days, and the heating repeated. This time the loss in weight amounted to 0.17 g or 0.16% of its weight, and a water content of about 0.00061 g/cm^2.

15. R. M. Organ, "The Safe Storage of Unstable Glass," *Museums Journal*, 56, February, 1957, pp. 265–272.

16. We thank Dr. Yao Su and Theresa Magliocca of Corning Glass Works for carrying out this determination.

Reequilibration Times

For the system being tested to be used successfully, it is necessary not only that the silica gel release or absorb water as required, but also that it do so quickly enough so as to prevent the RH from wandering outside the safe levels specified. In order to evaluate the gel in this respect, a few simple experiments were conducted in laboratory desiccator chambers.

The capacity of the chambers used was 9.7 l. Approximately 90 g of silica gel removed from the enamel trays of the experimental case was placed in a covered glass dish inside the chamber. Then either a small quantity of freshly dried silica gel desiccant or a small quantity of liquid water was placed alongside the dish. The chamber was closed and allowed to dry out or build up a high RH. The chamber was then opened slightly and the dish of equilibrated silica gel was uncovered and the desiccant or any remaining water removed. Thus, the uncovered dish of gel was exposed suddenly to rather high or low RH conditions, and the time it required to restore the RH to its own equilibrium RH was recorded by a sensor probe enclosed in the chamber. In both cases the gel responded quickly.

In one experiment, the gel adjusted a starting RH of 69% to 41% in about two hours. In another experiment, it adjusted a starting RH of 13% to 40% in about four hours. It is probably true that a small-volume system reequilibrates more efficiently than a large system, but since the ratio of the chamber volume to exposed gel area was about the same in these experiments as it was in the large experimental case, it was surmised that the recovery time of the large case must be well within acceptable limits.

An experiment on the Crizzling Case itself showed that when the case was flushed out with ambient air at 17% RH, the silica gel reestablished its equilibrium RH of 40% within about eight hours. (The increase to 35% RH, 75 per cent of the way, required only two hours.)

A Special Precaution

Usually, when water had to be added to the system, this was done by removing the trays of silica gel and spraying or squirting a measured volume of distilled and deionized water onto them. On a few occasions, however, liquid water was added by placing it in twelve small beakers distributed throughout the case. In these instances, the RH rose temporarily, but not excessively, because the water evaporated more quickly than the gel could take it up. No harmful effects were noticed until the third time this was done. On that occasion, some water was sprayed onto the gel and some placed in the beakers. The moistened gel was evidently even slower in taking up water vapor from the air. The RH rose to a level of 70–80% and stayed in that range for about fifteen hours. At that time, it was noticed that one of the objects, a Venetian goblet (CMG no. 59.3.20), had become separated at an old repair and collapsed sideways. Fortunately, it did not fall, for if it had, it certainly would have broken badly, and could have taken other objects with it. Upon examining the rest of the objects in the case, it was found that two others had similarly become unglued. No real harm was done, because all of the separations were at old breaks. We mention this not only as a precaution against adding water to a system in this way, but even more importantly, as a warning that edge-to-edge joins made with epoxy resins, as these had been, can apparently be loosened by rather short-term exposures to an RH of the order of 70–80%. This was totally unexpected by us and should be noted very carefully by all those who handle repaired glasses.

Short-term Variations in Relative Humidity

Careful observations were made of small variations in RH which occurred over a matter of a few hours or a day. Normally, these variations were no greater than one or two percentage units of RH, but a few changes as great as three or four percentage units were observed. Generally these larger changes were accompanied by changes in temperature. When a temperature change of five or six degrees occurred, which occasionally happened, the RH would usually change correspondingly; that is, a rise in temperature would cause a decrease in RH until the silica gel could reequilibrate the system. But sometimes a rise in temperature was associated with a rise in RH, as if one were watching a classical adsorption experiment, in which the substrate releases the adsorbate as the temperature rises. These are small but interesting effects, and the processes should be studied further. In this connection, the fact that the RH of a given silica gel is thought to be relatively insensitive to temperature changes implies that the equilibrium partial pressure of water vapor over the gel must rise with temperature at about the same rate as the vapor pressure of water does.

Storage at Low Relative Humidity

In our earlier research several experiments were carried out on a Silesian goblet dated to 1820 which was donated for experimental purposes by Mr. Jerome Strauss.[17] The object, sample no. 4011, showed typical incipient crizzling. Chemical analyses were made of the glass, an electron microbeam probe profile analysis was made of the surface, and several experimental pieces were sawed from the rim and foot.

After the experiments were completed, a few of the sawed pieces were placed in chambers with controlled relative humidities. As reported earlier, when stored overnight at near-zero RH, the glass degenerated dramatically to a severely crizzled state. But it was not then known how low the RH could go without causing degeneration. We now have some further information on that subject. The chamber with the next lowest RH had a level of 12–14%, which was established over a saturated solution of lithium chloride. Before the fragment of glass was placed in this chamber, a series of photomicrographs was taken of its surface. The glass was placed in the chamber on January 31, 1975, and removed on October 25, 1977. (A few casual examinations had been made in the interim.)

It appeared to the author that this glass sample showed somewhat more prominent crizzling after the two years and nine months storage at the low RH than it did at the beginning of the experiment. A microscopic examination was made, and the results were compared to the photomicrographs taken at the beginning. Most of the fissures which could be seen after the storage at low RH were also visible in the original photomicrographs, there being only about a 2–3% increase in the number of fissures. This relatively small increase in the number of fissures would not account for the apparent degeneration toward severe crizzling, but a microscopic comparison between this sample and another sample stored at 70% RH did provide evidence which could explain it. The fissures in the glass stored at the lower RH were noticeably more prominent. They had the appearance of being wider and "more open" than those on the glass which had been stored at the higher RH. Therefore, although storage at the low RH had not appreciably increased the number of fissures in the surface, it had opened them up. Consequently, one must conclude that storage at an RH in the range of 12–14% for two to three years promotes a degree of dehydration, which is dangerous for glass with incipient crizzling.

A Comment on Treatment

Although the corrosion of glasses has been extensively studied, there is not complete agreement upon the details of the chemical mechanism by which the attack of water proceeds.

---

17. Sample no. 4011, Figure 3, in reference cited in footnote one.

The mechanism may be complicated and indeed may vary from situation to situation and from one glass system to another. Some knowledge of the details of the mechanisms (or, at least, a working hypothesis about them) is always necessary when attempting to devise treatments for various types of early glasses. This is true for stained glasses and ancient glasses as well as for crizzled glass objects. We have some thoughts on the matter and some experiments to suggest, but, by and large, details of the mechanism can be established and accepted by the community of glass scientists only after a great many experiments on many different glasses under varying and carefully-controlled conditions have been carried out. Such an undertaking is well beyond the scope of our Museum's scientific program, but we wholeheartedly encourage others to pursue such studies.

One point upon which there is general agreement is that the hydrolysis of glasses, their chemical attack by water, results in a depletion of alkali at the affected surface of the glass, and the conversion of the glass into a hydrated state. Numerous experiments, including several done on crizzled objects at our Museum, have confirmed this depletion of alkali.

An important point we wish to make here is in response to a suggestion which has frequently been raised, to the effect that if crizzling is caused by the removal of alkali, why not simply replace the alkali to reverse the process and restore the glass? The problem is that when the alkali is removed and the glass becomes hydrated, the molecular structure of the glass is altered. That is to say, its silicate network, or the "skeleton" of the glass, is also almost certainly disintegrated by the same process that causes the leaching of the alkali. The resulting structure is probably more like that of a hydrated silica gel than that of the original glass. Therefore, simple replacement of the alkali to its original concentration level should not be expected to transform the hydrated material back into a glass any more than the addition of carbon dioxide and water to a piece of charcoal should be expected to transform it back into an oak log. It is the way atoms are bound together—as much as what atoms are present—which determines the physical nature of matter, in this case the glassy character. Therefore, there is little hope, in the author's opinion, that the process of crizzling can be reversed chemically without some high-temperature process which would fuse the surface of the material and reconstitute it into a glass. Because high-temperature treatments are not likely to prove both feasible and effective, we are left only with applications of coatings or impregnations to consolidate a crizzled surface into something resembling its original glassy state.

Until now, most efforts at consolidation have proven unsuccessful in that the resulting appearances of the treated objects have been, if not displeasing, at least not glasslike. It may be that recent research at The Corning Museum of Glass[18] will offer some hope in this regard. Pretreatment of other types of deteriorated glasses with silanes has improved adhesion and penetration of consolidants. Such treatments may allow greater latitude in the selection of consolidants and their means of application to crizzled glasses, so that appearances can be achieved which more closely resemble the original glass surfaces.

A Final Note

As this article was going to press, we had a most informative discussion with Dr. Stolow regarding the construction of a new case to be installed in our existing gallery. This case, which will contain ancient glass, will be controlled with silica gel and will be used as an

18. Unpublished work by Raymond Errett of the Museum staff and Shelley Reisman, then an intern at the Museum, in consultation with Dr. Merrill Lynn of Corning Glass Works.

experimental installation for the new building. Dr. Stolow made many very helpful suggestions and comments, some of which have a direct bearing on the contents of this article. Readers interested in this subject should watch for his forthcoming publications in the conservation literature.

It will be recalled that in several instances we added water to silica gel by spraying or misting it directly onto the surface with raking. Dr. Stolow does not recommend this procedure for several reasons. First, it can lead to decrepitation of the gel, that is, the gel may break up into smaller particles. This can be a disadvantage in systems where the gel is supported in fabric-covered panels, as the smaller particles fall out more easily; and also, small particles produced by decrepitation may cause dusting. Perhaps the most serious objection to adding liquid water directly to the gel is that it is difficult to add the water uniformly—and this difficulty becomes accentuated when one has to deal with large quantities of gel. In a paper presented at the meeting of the American Institute for Conservation in Fort Worth during June, 1978, Dr. Stolow discussed systems for conditioning silica gel in controlled-humidity chambers.

Dr. Stolow feels that the presence of other water-abundant phases in a system controlled by silica gel can have a beneficial effect by helping the gel to buffer the system, even though it does introduce another variable. He also prefers the use of larger-sized particles in order to prevent their escaping through the fabric covers of panels supporting the gel. Because our trial installation will involve the use of such panels, a mesh size of 3–8 will be used in accordance with Dr. Stolow's recommendation.

Acknowledgements

It would be difficult to list here the names of all those who contributed to this complicated project. The author hopes that those whose names are not mentioned below will accept his thanks anonymously and take satisfaction in knowing their efforts were directed toward a worthwhile cause. Special thanks are due Judy Snyder, who for three years recorded the data and looked after the experimental cases. She also assisted in the preparation of the manuscript. In the later stages of this work, Susan Johnson ably carried out these same responsibilities. Dr. David J. Fischer conducted several preliminary studies which, although they are not discussed in the text, were essential to the project. He was assisted by Joseph Maio, who also helped in maintaining the experimental cases. All too often the efforts of members of museum security staffs go unnoticed, so it is a special pleasure to acknowledge here the willing cooperation of the guards who took data during those times the Museum was closed. They are: William Buchko, Patrick Eldred, Charles Gardner, Harold Gray, Ronald Ogden, Kerry O'Shanick, Walter Schroeder, Deborah Tally, Joseph Turchick, Roswald Viele, and Paul Yorio.

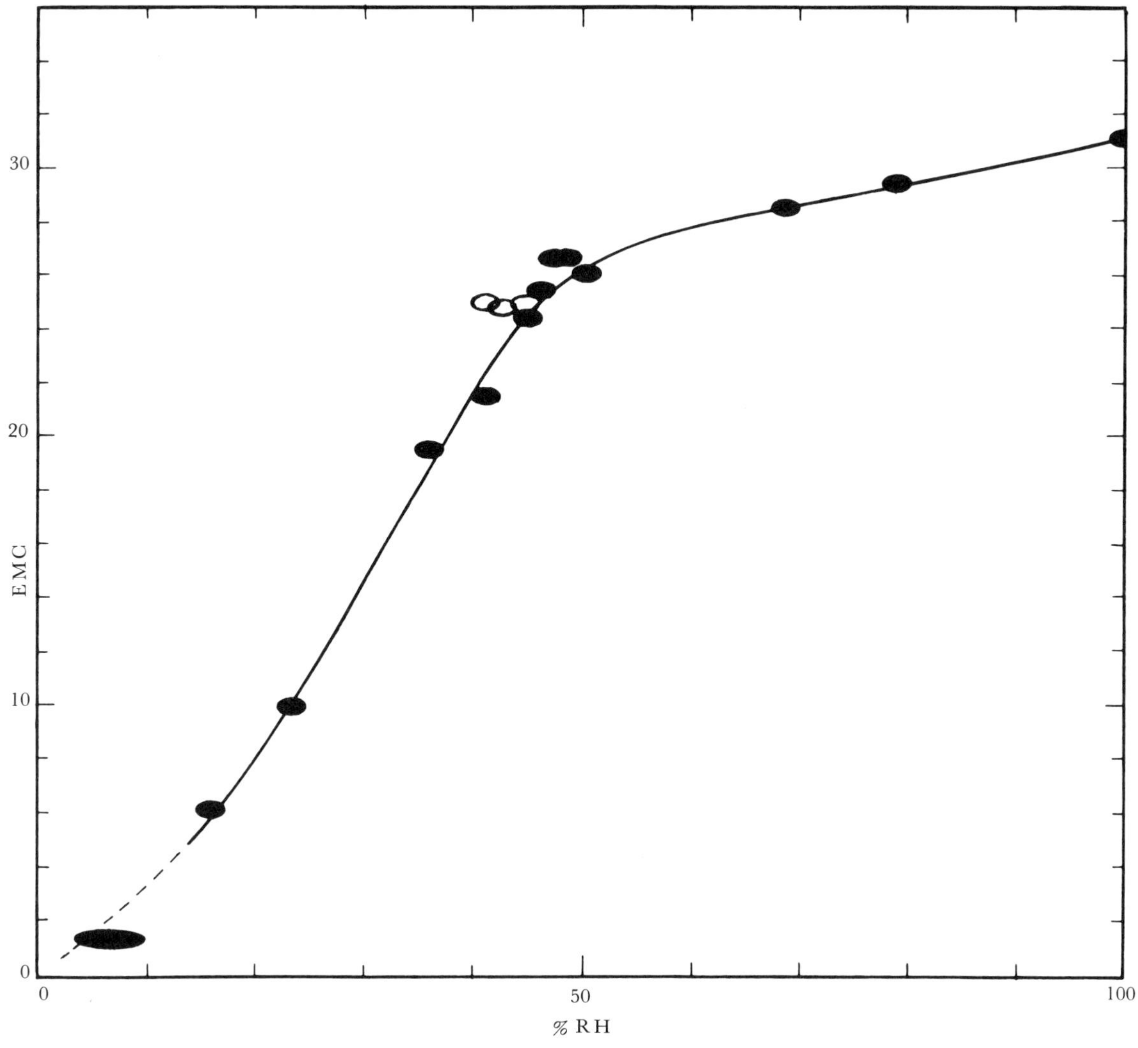

FIG. 1. *Equilibration curve for silica gel used in the Crizzling Case experiment. The curve shows the equilibrium RH of air in contact with prepared samples of silica-gel having various moisture contents. (EMC is expressed as percentage of water relative to dry weight of gel.) Hysteresis is not indicated because of the small number of points determined. Open ellipses are data for samples of gel removed from the case at three stages of the experiment.*

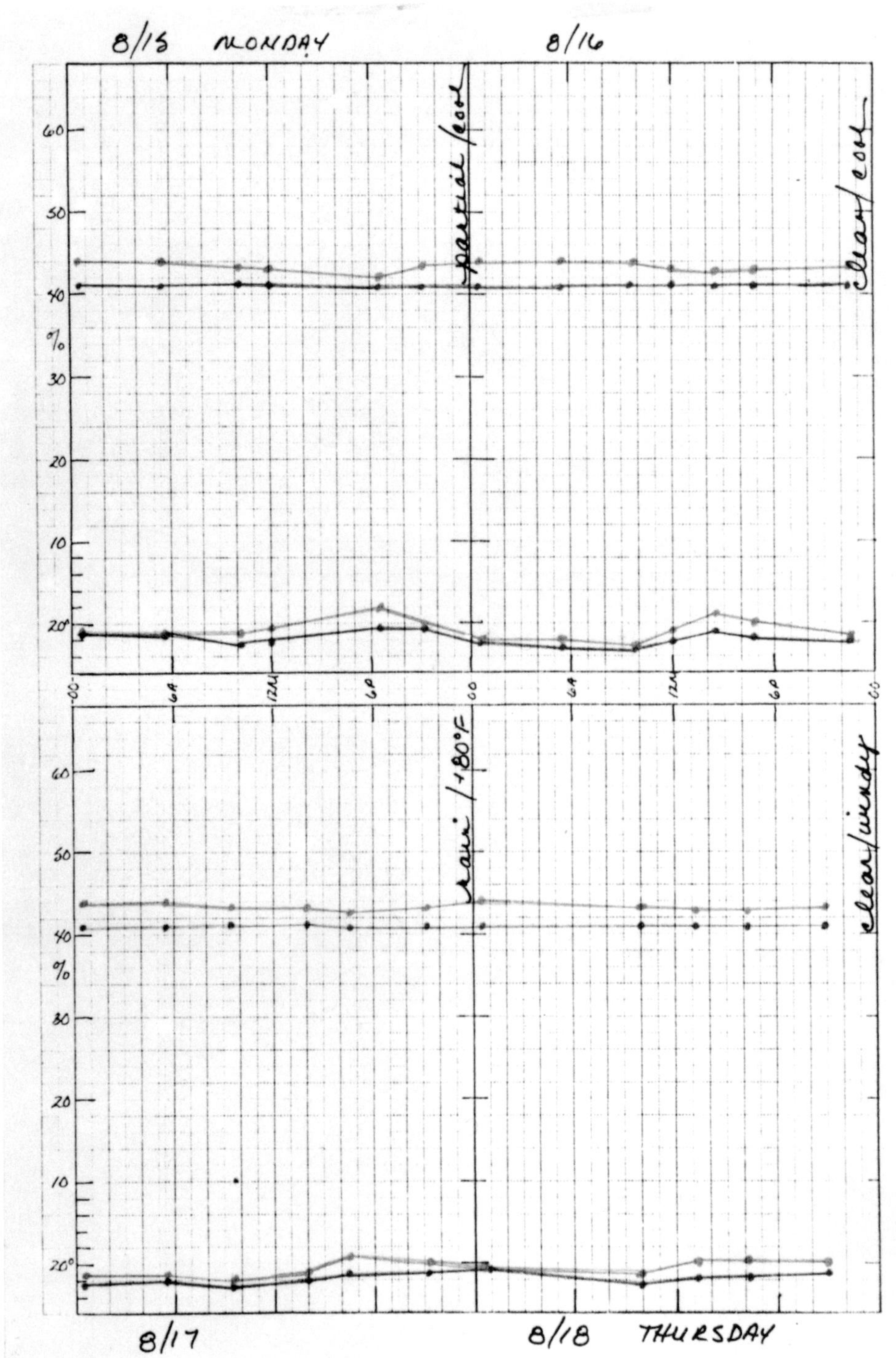

FIG. 2. *Page from Mrs. Snyder's and Mrs. Johnson's daily log book. Data are for four mid-summer days in 1977 and are typical of periods when the Crizzling Case was well stabilized. The plot shows raw data, not yet corrected for a slight calibration difference between the two probes.*

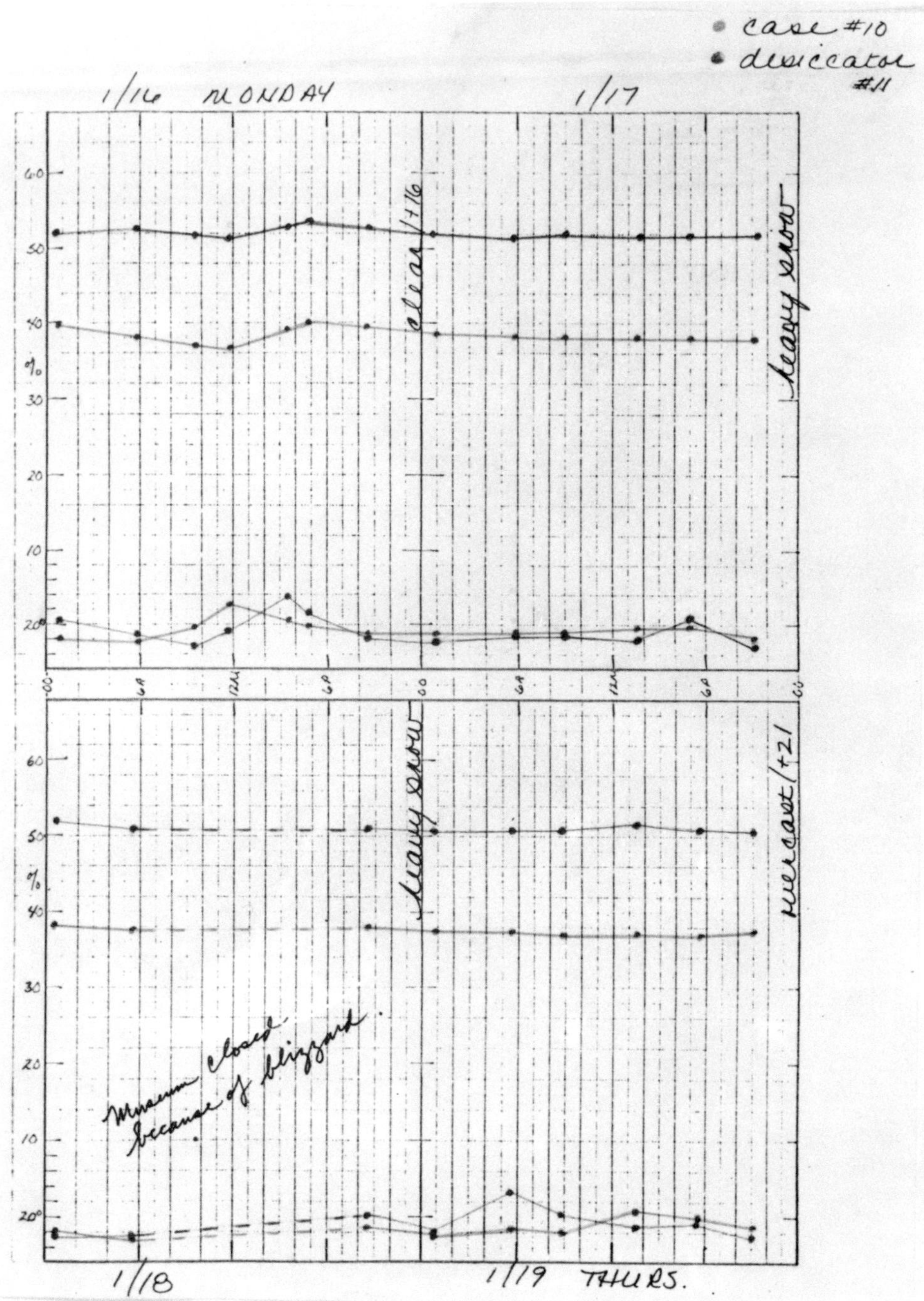

FIG. 3. *Page from Mrs. Snyder's and Mrs. Johnson's log book. Data are for four mid-winter days early in 1978 when there was a heavy snowfall. The lower of each pair of RH curves is for the Crizzling Case while it was drying out exponentially. The upper RH curve is for one of the chambers of silica gel used in establishing the RH* versus *EMC curve (EMC = 26.1%), which was just approaching its equilibrium state. The peaks in the temperature curves at the bottoms of the graphs result from occasional increases in the ambient temperature owing to some peculiarity of the building's heating system.*

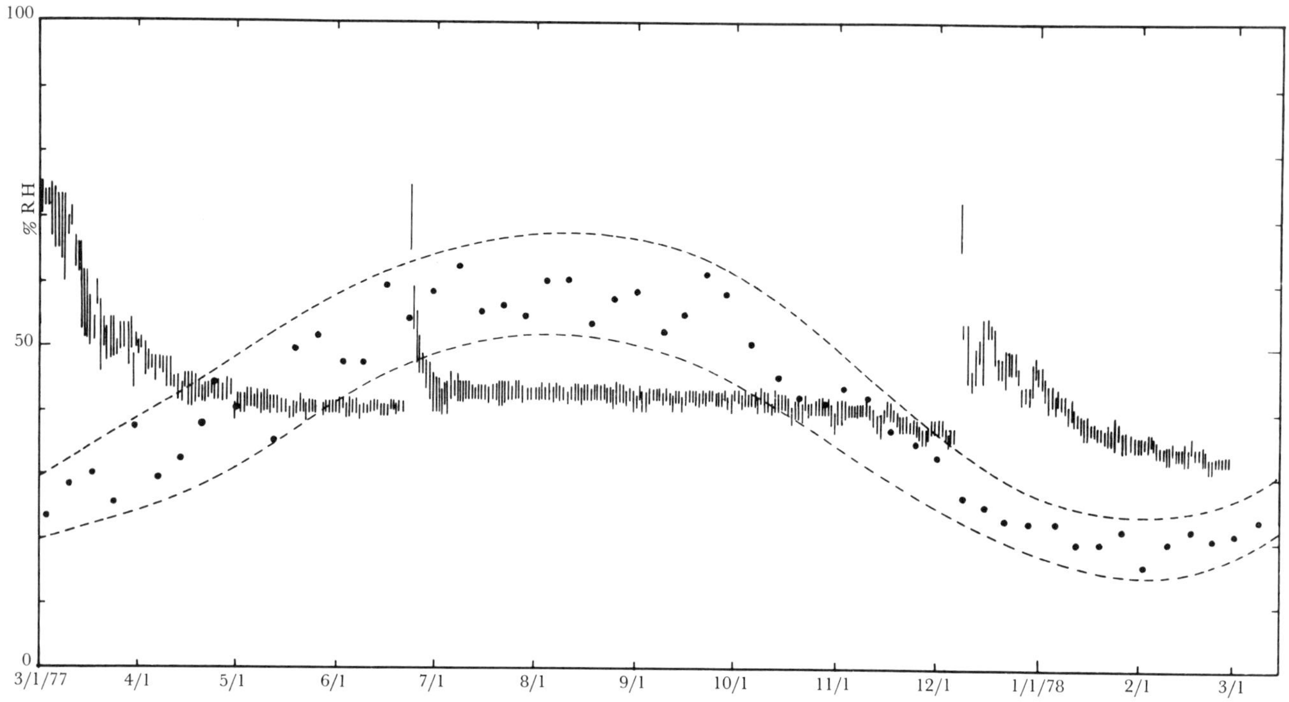

FIG. 4. *One-year RH history of the experimental Crizzling Case plotted as a daily record. RH values measured at two locations within the case are plotted as vertical lines. The length of each line represents the daily range. Additions of water to the system are evident in early March 1977, late June 1977, and early December 1978. In all, some 5000 measurements were used to construct this curve. Solid dots indicate weekly mean ambient RH values immediately outside the case. The broken curves are RH limits which include approximately 95% of the recorded data. The central tendency of RH values measured in a different part of the Museum, the Ancient Gallery, generally falls within the same limits, but the RH tends to wander outside the limits more frequently to temporary extremes of higher or lower RH.*

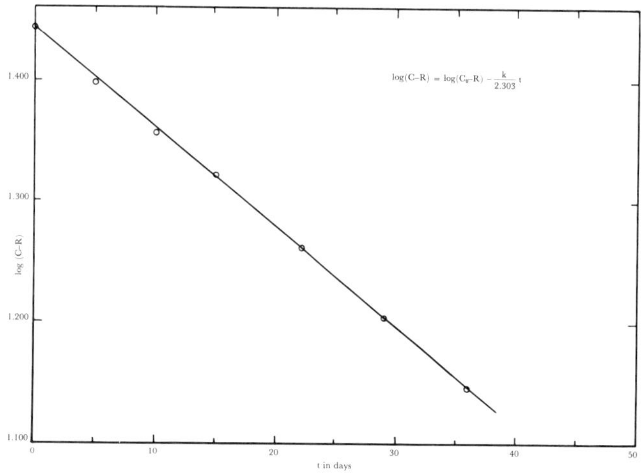

FIG. 5. *Plot of log (C-R)* versus *time for RH data recorded in the Crizzling Case experiment during the interval December 17, 1977 till January 23, 1978. C is the RH measured within the case at time t, R is the average ambient RH, which is known to have been about 22%, and C_0 is the RH within the case at t = 0. The resulting straight line indicates that the drying out of the case is exponential with time, in accordance with the model proposed by Thomson. From the slope of the line, a rate constant of k = 0.0192 per day was calculated, which yields a hygrometric half-time of 36.1 days.*

Recent Important Acquisitions

MADE BY PUBLIC AND PRIVATE COLLECTIONS IN THE UNITED STATES AND ABROAD

1. Gold-glass bowl fragment, colorless, gold-glass technique. Hellenistic, 3rd–2nd century B.C. H. (restored) 8.4 cm. *Museum of Art and Archaeology, Columbia, Missouri* (No. 77.198).

2. Gold-band alabastron, recut in antiquity. Roman Empire, late 1st century B.C.–1st century A.D. H. 7.5 cm. *Harvey Rothenburg, New York.*

3. Millefiori bowl, light purple matrix, mosaic glass. Eastern Mediterranean, late 1st century B.C.–1st century A.D. H. 4.2 cm. *Indiana University Art Museum, Bloomington, Indiana* (No. 76.35.70; Burton Y. Berry Collection).

4. Bowl, yellow-brown, blown, cut. Roman Empire, 1st century B.C.–1st century A.D. OH. 5.0 cm. *Erwin Oppenländer, Waiblingen, Germany* (No. 22652).

5. Flask with fruit motifs, opaque white, mold-blown. Roman Empire, "Sidonian," 1st century A.D. H. 8.3 cm. *Harvey Rothenburg, New York.*

6. Double-headed beaker, bright green, mold-blown. Roman Empire, Eastern Mediterranean, 1st century A.D. H. 8.7 cm. *Erwin Oppenländer, Waiblingen, Germany* (No. 22345).

1

2

3

6

4

5

7

8

9a

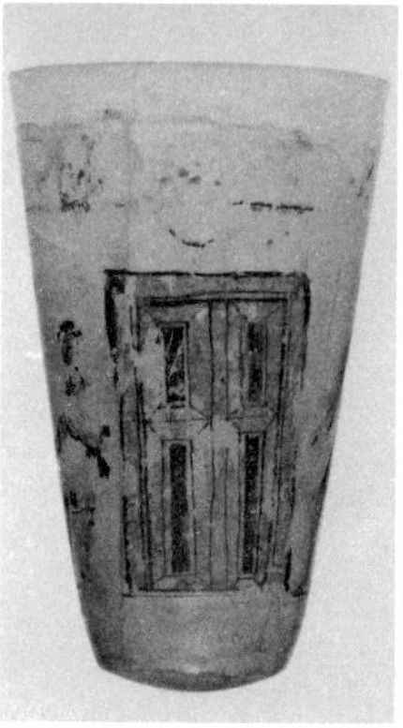

9b

7. Seven dining pieces from a tomb group, green, blown. Roman Empire, 1st century A.D. D. (plate) 17.9 cm. *The Corning Museum of Glass* (No. 77.1.2 A–G).

8. Cinerary urn, pale green, blown. Romano-German, from Krefeld-Gellup, Germany, late 1st–2nd century A.D. OH. 32 cm. *The Toledo Museum of Art, Toledo, Ohio* (No. 77.14; gift of Edward Drummond Libbey).

9. Beaker, colorless, with enameled decoration and Greek inscription, blown and cut. Eastern Roman Empire, 3rd century A.D. H. 14.5 cm. *Hans Cohn, Los Angeles, California.*

10. Jug with woven covering, green, blown. Roman Empire, Egypt, 4th century A.D. H. 25.2 cm. *The Corning Museum of Glass* (No. 77.1.3).

11. Pitcher, blown-molded. Syro-Palestinian, ca. 600 A.D. H. 15.8 cm. *Hans Cohn, Los Angeles, California* (No. 211).

10

11

12. Palm cup, blown. Frankish, excavated near Bingen, Germany, 6th–7th century A.D. H. 7 cm. *The Toledo Museum of Art, Toledo, Ohio* (No. 77.11; gift of Edward Drummond Libbey).

13. Basket, greenish, blown, with pinched decoration. Found in Nishapur, northeast Iran, 9th–10th century A.D. H. 8.9 cm. *Hans Cohn, Los Angeles, California* (No. 208).

14. Jug, green, blown, with pinched decoration. Near East, possibly Egypt, 8th–10th century A.D. H. 6.7 cm. *The Indiana University Art Museum, Bloomington, Indiana* (No. 76.35.72; Burton Y. Berry Collection).

15. Plate, colorless, blown and cut. Found in Nishapur, northeast Iran, 10th century A.D. D. 23.3 cm. *Hans Cohn, Los Angeles, California* (No. 201).

16. Bottle, purple with gilt decoration, blown. Byzantine, late 12th century. H. 15.9 cm. *The British Museum, Department of Medieval and Later Antiquities, London* (No. 1977.77-1, 1).

17. Beaker, colorless with transparent ruby threads, blown, twelve cut and engraved panels. Bohemia, ca. 1700. OH. 20.1 cm. *The Corning Museum of Glass* (No. 77.3.8; gift of Mrs. K.F. Landegger).

12

13

14

15

16

17

18

19

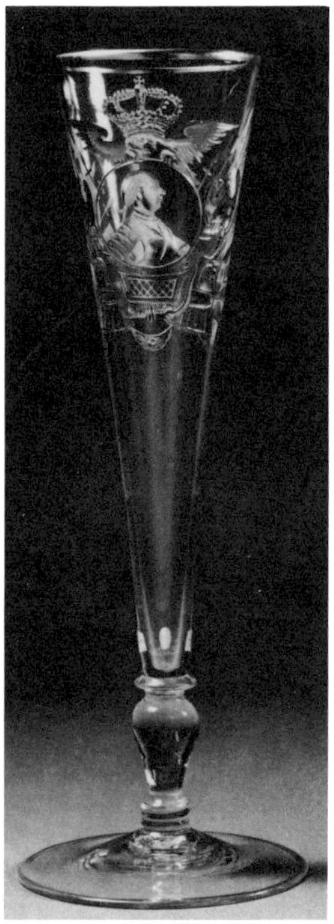

20

21

18. Small goblet, colorless, blown, engraved portrait of Charles VI and arms of the German Empire. Nuremburg, ca. 1711–1712, probably engraved by H. W. Schmidt. *Rudolf von Strasser, Pelham Manor, New York* (No. G 257).

19. Beaker, colorless, blown, engraved arms of the Elector of Bavaria. Possibly Bohemia, ca. 1720. H. 8.7 cm. *Bayerisches Nationalmuseum, Munich* (No. 76/114; gift of Emil Seidl).

20. Flute glass, colorless, blown, engraved and gilt portrait of Friedrich Wilhelm I of Prussia. Potsdam-Berlin, ca. 1730. H. 30.5 cm. *Elisabeth Osthoff, Bielefeld, Germany.*

21. Goblet, colorless, blown, cut, engraved *hochschnitt* arms of Georg Wilhelm Von-Schwei nischen, and a boar hunt in *tiefschnitt*; (replaced) gilt-copper foot. Silesia, dated 1747. OH. 27.3 cm. *The Corning Museum of Glass* (No. 77.3.10; gift of Mrs. K. F. Landegger).

22. Beaker, colorless, blown, cut, molded sulphide portrait of Louis XVIII. France, Paris, Desprez, or Boudon de Saint-Amans; initials of Franz v. Pillement on the sulphide, ca. 1820. H. 9.6 cm. *Bayerisches Nationalmuseum, Munich* (No. 77/37).

22a

22b

23. Dish, opaque yellow, carved decoration. China, possibly Peking, mark and period of Ch'ien-lung (1736–1796). D. 13.8 cm. *Royal Ontario Museum, Far Eastern Department, Toronto* (No. 976.316.7; gift of Mr. R.G. Nicholas Laidlaw, Mrs. R.W.L. Laidlaw, and Mrs. Andrew Smith).

24. Chandelier, colorless, blown, cut. England, ca. 1770. OH. 94 cm. *Colonial Williamsburg, Williamsburg, Virginia* (No. 1977–37).

25. Tumbler, colorless, blown, engraved. United States, probably New Bremen Glassmanufactory, Maryland, ca. 1785–1795, possibly 1788–1789. H. 15 cm. *Frank T. Innes, Chester, New Jersey.*

26. Decanter, green, blown-molded. United States, probably Mantua Glass Works, Mantua, Ohio, ca. 1825. H. 21 cm. *Duncan B. Wolcott, Cuyahoga Falls, Ohio* (No. 106).

27. Pitcher, pale olive green, blown. United States, attributed to the Rockville Glass Manufactory, Steubenville, Ohio, ca. 1835–1846. H. 9.9 cm. *Royal Ontario Museum, Toronto* (No. 977.142; gift of Rudy Eswarin).

23

24

25

26

27

28

29

30

31

28. Jug, colorless, blown, enameled. England, decoration by Richard Redgrave, ca. 1850. H. 26.6 cm. *The Cleveland Museum of Art, Cleveland, Ohio* (No. 77.76).

29. Goblet, colorless and dark blue, engraved. Bohemia, engraved by F. Zach (signed), ca. 1850–1860. *Rudolf von Strasser, Pelham Manor, New York* (No. G 255).

30. "Ludion" Cartesian Diver, colorless, blown, enclosing water and colored lampworked figures symbolic of the crucifixion and animals floating inside. Europe, probably France, mid-19th century. H. 36.0 cm. *The Corning Museum of Glass* (No. 77.3.25; gift of the Comtesse J. de Vogüé).

31. Vase, purple and green with silver-gilt base, blown, cameo-carved. Paris, possibly by Edward Michel, ca. 1900. OH. 21.5 cm. *The Denver Art Museum, Denver, Colorado* (No. 1977.56).

32. Footed bowl, aquamarine, blown. New York, Louis C. Tiffany, ca. 1900. H. 19.2 cm. *The Cooper-Hewitt Museum of Decorative Arts and Design, Smithsonian Institution, New York* (No. 1975-32-14).

33. Bowl, marbleized violet, bronze mounts, blown. France, Daum, Nancy; mounts by L. Majorelle, ca. 1930. H. 18 cm. *Musée du Verre, Liège.*

34. Vase, blown. France, Henri Navarre, ca. 1930–1935. H. 22.5 cm. *The Cleveland Museum of Art, Cleveland, Ohio* (No. 77.12).

32

33

34

35. Vase, colorless, blown and engraved. Steuben Glass, Corning, New York, ca. 1932. H. 18.5 cm. *Yale University Art Gallery, New Haven, Connecticut* (No. 1977.36.1; gift of Mrs. H. Emerson Tuttle).

36. Pitcher and four glasses, colorless, blown and cut. J. & L. Lobmeyr, Vienna, designed by Adolf Loos in 1934. H. (pitcher) 23.9 cm. *The Corning Museum of Glass* (No. 77.3.36 A–E; gift of J. & L. Lobmeyr).

37. The "Valor Cup," colorless, blown, engraved. Steuben Glass, Inc., Corning, New York. A duplicate of the original, made for the British War Relief Society to commemorate the Battle of Britain; designed by John Monteith Gates, made by John Jansson, engraved by Joseph Libisch, 1940. OH. 39.8 cm. *The Corning Museum of Glass* (No. 77.4.90; gift of John Monteith Gates).

38. Cross and candlesticks, colorless, cast by the *cire perdue* process. Steuben Glass, Inc., Corning, New York, designed by Paul V. Gardner, 1941. OH. (cross) 53.2 cm. *The Corning Museum of Glass* (No. 77.4.87 A–D; gift of William H. H. Hower, II).

39. "Etude," yellowish, cast. Leningrad, Russia, Department of Glass, Leningrad Institute of Technology, by Vera Muchina and Nikolai Kachalov, 1947, H. 13.9 cm; L. 24 cm. *The Corning Museum of Glass* (No. 77.3.45; gift of the Ministry of Foreign Trade and the Ministry of Electronic Industry, U.S.S.R.).

37

38

35

36

39

40

41

42

43

40. "Red Flower" or "Two Hearts," red, cast. Stanislav Libenský and Jaroslava Brychtová, Železný Brod, Czechoslovakia, 1974. H. 85.3 cm. *Museum Bellerive, Zurich* (No. 1977–15).

41. "Three Disks," optical glass, cut. Václav Cigler, Prague, Czechoslovakia, designed in 1969, executed in 1976. H. 27 cm. *Museum Bellerive, Zurich* (No. 1977–18).

42. "Steuben Prototype #17," iridescent, blown. United States, designed and made by Dale Chihuly and James Carpenter, 1975. OH. 22 cm. *Wadsworth Atheneum, Hartford, Connecticut* (No. 1977.67).

43. "Madonna and Child," colorless, cut and engraved. Steuben Glass, Corning, New York, designed by Bernard X. Wolff from a sculpture by Bruno Lucchesi, interpreted and engraved by Lubomir Richter, 1976. H. 33.4 cm. The Corning Museum of Glass (No. 77.4.3; anonymous gift through the generosity of Mr. and Mrs. Edwin J. Beinecke, Jr.).

44. "A Movement of Change," colorless, cut, on aluminum base. England, Peter S. Aldridge, 1976. OH. 25.5 cm. *The Corning Museum of Glass* (No. 77.2.1).

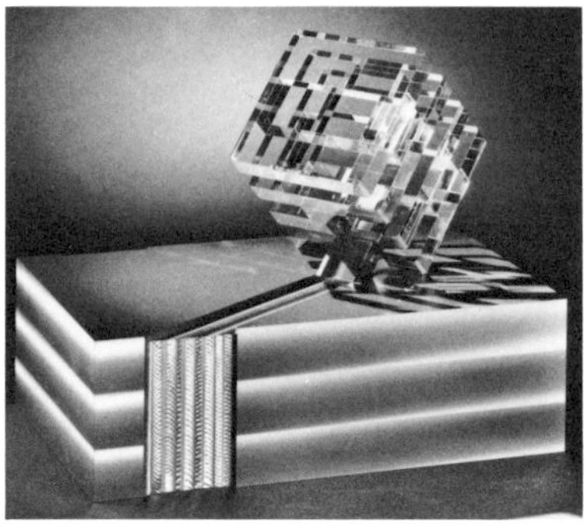

44

CHECK LIST

of Recently Published Articles and Books on Glass

1977

This list includes all publications added to The Corning Museum of Glass library since the Check List for *Journal* XIX. In some cases, older material new to the library collection has been added, excluding any article or book prior to 1940 listed in *Bibliography of Glass* by George S. Duncan, Sheffield: Society of Glass Technology, 1960.

INDEX

Fewer than five citations from any area are included under Renaissance to Present, General.

Only substantive book reviews are listed and may be found under the author of the work reviewed.

Collectibles are included under country of origin.

Articles on restoration and conservation of Stained Glass are listed under Technology.

GENERAL

ANONYMOUS

"A Close Look at the History of Eyeglasses"
PPG Products, PPG Industries, 85, 1977, pp. 8–9, illus.

"Glass and the Builder"
Monthly Bulletin for the Glass Industry, Cerglas Ltd., 503, June/July 1977, p. 1.
Glass as a building material.

"Glassammlungen der DDR"
Bulletin de l'Association Internationale pour l'Histoire du Verre, 7, 1973–1976, pp. 25–138, illus.

"Making New Laboratory Items Look Antique Presents Problems for Kimble Craftsman"
Owens-Illinois Outlook, No. 201, April 1977, p. 2, illus.
Lampworker recreates 18th c. laboratory pieces for Priestley House.

ALLISON, GRACE
"Toledo Museum of Art Glass Collection"
Rainbow Review, 7, No. 12, Dec. 1977, pp. 20–24, illus.

Antike Gläser der Cinzano-Sammlung
n.l.: Cinzano, n.d., 38 pp., illus.
Collection ranges from Roman-Islamic to 18th c. European.

BOLKEN, KANDYCE SMITH
"Glass Reproductions and Fakes"
Depression Glass Daze, 7, No. 4, June 1, 1977, p. 30.

BROOKS, JOHN
"Questioning Collector I: Glassware; John Brooks Advises the New Collector"
Antique Dealer and Collectors Guide, June 1977, pp. 128–130, illus.

CHILD, FRANK S.
"The Finer Points of Copper Wheel Engraving"
American Glass Review, 97, No. 9, March 1977, pp. 8–10+, illus.

EHRESMANN, DONALD
Applied and Decorative Arts; A Bibliographic Guide to Basic Reference Works, Histories, and Handbooks
Littleton, Colorado: Libraries Unlimited, Inc., 1977, 232 pp.
Selected bibliography of books on glass: dictionaries, handbooks, and glass of different periods and countries.

ENGLE, ANITA, ed.
Readings in Glass History, No. 9; Index to Nos. 1–8
Jerusalem: Phoenix Publications, 1977, 73 pp.

Glass the Miracle Worker
Toledo, Ohio: Libbey-Owens-Ford Co., [1977], 15 pp., illus.

GREEN, JANET
"What Do You See in Glass?"
Antique Collector, 1, No. 1, May 1977, pp. 8–9, illus.
Brief history.

HALLE. STAATLICHE GALERIE MORITZBURG
Glas aus zwei Jahrtausenden; Bestände der Galerie von 700 v.d.Zt. bis 1975
Halle: Staatliche Galerie Moritzburg, 1977, 68 pp., illus.

HOLLISTER, PAUL
"Glasshouse Working Conditions"
Glass Paperweight Symposium, June 1976, The John Nelson Bergstrom Art Center and Museum, Neenah, Wisconsin, pp. 48–53.

HOLLISTER, PAUL
"Through a Glass, Darkly"
The Glass Club Bulletin, 119, March 1977, pp. 4–11, illus.
History of window glass.

HOLSCHER, HARRY H.
The Historical Development of Lids, Stoppers and Closures and Their Effects upon the Design of the Container
Toledo, Ohio: Owens-Illinois, Inc., n.d., n.pag.
Xeroxed copy of a typewritten paper, illus.

KAELLGREN, C. PETER
"The Constancy of Glass"
Canadian Collector, 12, No. 4, July/August 1977, pp. 24–27, illus.

KAZIUN, CHARLES
"Lampwork Paperweight Designs"
Glass Paperweight Symposium, June 1976, The John Nelson Bergstrom Art Center and Museum, Neenah, Wisconsin, pp. 54–60.

KEEFE, JOHN
"The Care, Restoration, and Photography of Glass Paperweights"
Glass Paperweight Symposium, June 1976, The John Nelson Bergstrom Art Center and Museum, Neenah, Wisconsin, pp. 95–104.

LABINO, DOMINICK
"Some Important Properties of Glass"
Glass Paperweight Symposium, June 1976, The John Nelson Bergstrom Art Center and Museum, Neenah, Wisconsin, pp. 72–91, illus.

LAWRENCE, KANSAS. THE UNIVERSITY OF KANSAS MUSEUM OF ART
Sallie Casey Thayer and Her Collection
(text by Carol Shankel)
Lawrence, Kansas: The University of Kansas Museum of Art, 1976, 95 pp., illus.
Includes glass of all periods.

Master's Theses: Crafts; A National Directory Compiled by the American Crafts Council Research and Education Department
New York: American Crafts Council, 1976, 153 pp.
Glass, stained glass entries back to the 1950's.

NEWMAN, HAROLD
An Illustrated Dictionary of Glass
London: Thames and Hudson, 1977, 351 pp., illus.

PANCAKE, EVAN
"Objective: Paperweight Collecting"
Glass Paperweight Symposium, June 1976, The John Nelson Bergstrom Art Center and Museum, Neenah, Wisconsin, pp. 61–71, illus.

PHILIPPE, JOSEPH
Le Musée du Verre de Liège
Liège: Librarie Halbart, 1976, 24 pp., illus.

Pilkington Glass Museum; Evolution of Glassmaking Techniques: A Temporary Guide to the Collection
St. Helens: Pilkington Glass Museum, 1974, 22 pp.

POESE, BILL
Lighting through the Years
Des Moines, Iowa: Wallace-Homestead Book Co., 1976, 99 pp., illus.

PRAGUE. UMĚLECKOPRŮMYSLOVÉ MUZEUM
Acta Upm I. Přírůstky sbírek (Additions to the Collections 1965–1968)
Prague: Uměleckoprůmyslové Muzeum, 1969, n.pag., illus.
Glass of different periods, countries.

Ray, Helen and Raymond, and Keller, ed.
"Lampworking with Glass"
Fusion, 24, No. 1, Feb. 1977, pp. 9–16, illus.

Regensburg. Museum der Stadt
Gläser: Antike, Mittelalter, und Neuere Zeit
Karlsruhe: Corona-Verlag Karl H. Heine KG, 1977, 204 pp., illus.
Catalog of Museum der Stadt Regensburg and Sammlung Brauser.

Rotzler, Willy
"Glas im Spiegel der Kunst"
Du, 37, Aug. 1977, pp. 32–61, illus., English summary.
Representations of glass in paintings.

Rotzler, Willy
"Kleine Geschichte eines Werkstoffes durch die Jahrtausende"
Du, 37, Aug. 1977, pp. 20–31, illus., English summary.
General history of glass.

Toronto. The Royal Ontario Museum
A Gather of Glass; Glass through the Ages in the Royal Ontario Museum (edited by C. Peter Kaellgren)
Toronto: Royal Ontario Museum, 1977, 32 pp., illus.

TECHNOLOGY

Anonymous
"Chartres: Through a Glass Darkly"
Time, 109, No. 5, Jan. 31, 1977, pp. 54–55, illus.
Restoration issues.

"Consumer Report: Antique Glass"
Glass (formerly *Glass Art*), 5, No. 2, Feb. 1977, pp. 10–11, illus.
Analysis of contemporary handblown glass available to stained glass artists.

"Consumer Report: Domestic Opalescent and Cathedral Glass"
Glass (formerly *Glass Art*), 5, No. 1, Jan. 1977, pp. 42–43, illus.

"Corning Waveguide Work Is Exhibited at Smithsonian"
The Gaffer, 35, No. 3, April 1977, p. 4.

"La corrosion des vitraux"
Le Courrier du CNRS (Centre national de la recherche scientifique), No. 21, July 1976, pp. 29–33, illus.

"Pressed Glass I–XII"
Monthly Bulletin for the Glass Industry, Cerglas, No. 496, Nov. 1976–No. 507, Nov. 1977.
History of glass pressing.

Approaching Trends: Impact on Glass
Society of Glass Decorators Seminar, Pittsburgh, 1976, 148 pp., illus.
Papers presented at seminar, including "Glass; A Unique Medium" by Dominick Labino.

Bayer, Gerhard and Wiedemann, Hans-Georg
"Le bleu egyptien, pigment synthétique de l'antiquité examiné sous l'angle scientifique"
Bulletin Sandoz, No. 40, n.d., pp. 20–40, illus.

Becksmann, Rüdiger
"Probleme der Restaurierung spätromanischer Glasmalereien"
Kunstchronik, 29, 1976, pp. 330–337.
Late Romanesque stained glass restoration.

Bettembourg, Jean-Marie
"Problèmes de la conservation des vitraux de la façade occidentale de la cathédrale de Chartres"
Les Monuments Historiques de la France, 1, 1977, pp. 7–13, illus.

Billeci, Andre
Electric Melting Unit for Covered Melting
Corning: The Corning Museum of Glass, 1976, 16 typescript pp., illus.
Corning Museum Furnace Project.

Collongues, Robert
"La corrosion des vitraux"
Les Monuments Historiques de la France, 1, 1977, pp. 14–16, illus.

Cook, John
"Equipment and Safety"
Presented at *Working with Hot Glass*, international conference organized by the Crafts Advisory Committee and Royal College of Art, London, Sept. 1976.

Cummings, K. R.
"Non-Blown Techniques"
Presented at *Working with Hot Glass*, international conference organized by the Crafts Advisory Committee and Royal College of Art, London, Sept. 1976.
Millefiore, pillar-molding, etc.

Dauval'ter, A. N.
Khrustal'nye tsvetnye i opalovye stekla
(Crystal, Colored and Opal Glass)
Moscow: Gizlegproma, 1957, 235 pp., illus.
Compositions and properties.

Dean, R. R.
"Colour Patterned Glass by the Pulsed Electrofloat Process"
Tin and Its Uses, No. 108, 1976, pp. 7–8, illus.

Douglas, R. W.
"Redox Reactions and the Colours of Ancient Glasses"
Recent Advances in Science and Technology of Materials
Proceedings, Cairo Solid State Conference, 1973, (1974), pp. 47–59, illus.

Ferrazzini, Jean Claude
"Die Anwendung von Cyanoacrylat-Monomer-Klebstoffen in der Glasrestaurierung"
Information Bulletin, Verband der Museen der Schweiz, 16, June 1976, pp. 5–9.
Protection of stained glass with cyano-acrylate monomer adhesives.

Ferrazzini, Jean Claude
"Vorteile und Technik der Anwendung von Cyanoacrylat-Monomer-Klebstoffen zur Sicherung der Malerei von Glasgemälden"
Glastechnische Berichte, 49, No. 11, 1976, pp. 264–268, illus.
Advantages and techniques of application of cyanoacrylate monomer adhesives to protect the paint of stained glass.

Flavell, Ray
"Basic Glass"
Crafts, 27, July/Aug. 1977, pp. 22–27, illus.; 28, Sept./Oct. 1977, pp. 37–41, illus.
Techniques of glassblowing.

Flavell, Ray
"Glass Blowing from the Furnace"
Presented at *Working with Hot Glass*, international conference organized by the Crafts Advisory Committee and Royal College of Art, London, Sept. 1976.

Fleming, Stuart
Dating in Archaeology; A Guide to Scientific Techniques
New York: St. Martin's Press, 1976, 272 pp., illus.

FRODL-KRAFT, E.
"Mittelalterliche Glasmalerei; Erforschung-Restaurierung; Bemerkungen zu Verwitterungsformen und Konservierungsmassnahmen an mittelalterlichen Glasmalereien"
Österreichische Zeitschrift für Kunst und Denkmalpflege, 28, No. 4, 1974, pp. 200–209, illus.

FRODL-KRAFT, E.
"Mittelalterliche Glasmalerei; Erforschung-Restaurierung; Das. 9. Colloquium des Corpus Vitrearum, Paris, September 1975"
Österreichische Zeitschrift für Kunst und Denkmalpflege, 29, 1975, pp. 154–158, illus.

GLIKIN, S., SKROB, L., AND BONDARENKO, G.
"Primenenie svetnogo profil'nogo stekla v promyshlennoī arkhitekture" (Emploi de verre filtrant coloré dans l'industrie de la construction)
Arkitektura, 6, 1975, pp. 57–59, illus.

HLAVA, PAVEL
"Der Einzelne und die Hütte"
Kunst + Handwerk, 20, No. 10, Oct. 1976, pp. 365–366, illus.
Designer knowledge of blowing and glassworking techniques.

HUNT, L. B.
"The True Story of Purple of Cassius; The Birth of Gold-Based Glass and Enamel Colours"
Gold Bulletin (Johannesburg?) 9, No. 4, 1976, pp. 134–139, illus.

JAHN, ROLAND
"Energy Saving Glass Furnace-annealing Oven System: A Summary"
Presented at *Working with Hot Glass*, international conference organized by the Crafts Advisory Committee and Royal College of Art, London, Sept. 1976.

JENA LABORATORY GLASSWARE
Jena: Jena Glass Works Schott and Gen., 1937, 153 pp., illus.
Catalog of lab glass and apparatus.

KALLIO, HEIKKI
"Furnace and Annealer Designed by Heikki Kallio"
Presented at *Working with Hot Glass*, international conference organized by the Crafts Advisory Committee and Royal College of Art, London, Sept. 1976.

KNOTT, PETER
"Glass Furnace Technology"
Presented at *Working with Hot Glass*, international conference organized by the Crafts Advisory Committee and Royal College of Art, London, Sept. 1976.

KOKOCIŃSKA, KRYSTYNA AND SOCZEWIŃSKI, EDWARD
"Uwagi na temat metod i doświadczén pkz w zakresie konserwacji witrazy" (Remarques sur les méthodes et les expériences de l'atelier de conservation des monuments historiques dans le domaine de la conservation des vitraux.)
Ochrona Zabytków, 21, No. 3, 1968, pp. 31–40, illus., French summary.

LAWRIE, JOHN
"Studio Glass Decoration"
Presented at *Working with Hot Glass*, international conference organized by the Crafts Advisory Committee and Royal College of Art, London, Sept. 1976.
Cutting, polishing, copper-wheel engraving in a glass studio.

MARTIN, JOHN H., ed.
The Corning Flood: Museum under Water
Corning, N.Y.: The Corning Museum of Glass, 1977, 60 pp., illus.
Glass restoration.

MILLS, HARRY N.
"Raw Materials for Glass Manufacture"
The Glass Industry, 58, No. 11, Nov. 1977, pp. 10–15+, illus.

NÖLLE, GÜNTHER
"Grundprobleme der Formgebung des Glases"
Silikattechnik, 27, No. 8, 1976, pp. 263–266, illus.

O'KEEFE, JOHN
Tektites and Their Origin
Amsterdam: Elsevier Scientific Publishing Co., 1976, Developments in Petrology 4, 254 pp., illus.

PAWŁOWSKA, HANNA
"Die Fertigung von Eichproben und Spektralanalyse von Glasantiquitäten"
Archaeologia Polona, 17, 1976, pp. 75–87, illus.

PILKINGTON, ALASTAIR
"Flat Glass: Evolution and Revolution over 60 Years"
Glass Technology, 17, No. 5, Oct. 1976, pp. 182–193, illus.

PILTZER, WILLIAM AND BURT, ARTHUR
"On Photographing Glass"
Glass (formerly *Glass Art*), 5, No. 6, Oct. 1977, pp. 47–49, illus.

PRINDLE, WILLIAM A.
"Unusual Hot Glass Techniques: Color Encasement"
Glass (formerly *Glass Art*), 5, No. 1, Jan. 1977, p. 35, illus.

PRINDLE, WILLIAM A.
"Unusual Hot Glass Techniques: Hot Additions"
Glass (formerly *Glass Art*), 5, No. 2, Feb. 1977, p. 18, illus.

REBOULLEAU, MAGNIER
Nouveau manuel complet de la peinture sur verre, sur porcelaine et sur email (reprint)
Paris: Leonce Laget, 1977 reprint of 1866 edition, 419 pp., illus.

REHMATULLAH, CH.
"Use of Synthetic Resins for the Restoration of Glass Objects"
Pakistan Archaeology, 7, 1970/71, pp. 156–157.

SPITZER-ARONSON, MARTHA
"La distribution du cuivre dans les verres rouges des vitraux médiévaux"
C. R. Academie des Sciences Paris, Serie C: Sciences Chimiques, 278, June 17, 1974, pp. 1437–1440, illus.

STÄDLER, LEO AND CRONIN, DAVID
"Container Glass Composition"
The Glass Industry, 58, No. 12, Dec. 1977, pp. 10–14, illus.

STÄDTLER, LUDWIG
"Physikalische Grundlagen für die Laser-Applikation zur Bearbeitung von Wirtschaftsglas"
Silikattechnik, 27, No. 8, 1976, pp. 276–280, illus.

STANWORTH, J. E.
"Some Aspects of the Development of Glass Technology over 60 Years"
Glass Technology, 17, No. 5, Oct. 1976, pp. 194–204.

TARALON, JEAN
"Le colloque international d'Erfurt et la sauvegarde des vitraux anciens"
Les Monuments Historiques de la France, 4, Oct.–Dec., 1962, pp. 231–246, illus.

Taralon, Jean
"Problématique de la conservation et de la restauration des vitraux"
Les Monuments Historiques de la France, 1, 1977, pp. 2–6+, illus.

Taylor, R. E., ed.
Advances in Obsidian Glass Studies
Park Ridge, N.J.: Noyes Press, 1976, 360 pp., illus.

Uniprojekt: Designing, Engineering and Consulting Organization
Prague: State Printing Works, 1977, n.pag., illus.
Czech glass industry organization.

Valkema, Durk
"Furnace Developments at the Glass Workshop of the Gerrit Rietveld Academie, Amsterdam"
Presented at *Working with Hot Glass*, international conference organized by the Crafts Advisory Committee and Royal College of Art, London, Sept. 1976.

Weeden, Cyril
"The Glass Container Industry 1916 to 1976"
Glass Technology, 17, No. 5, Oct. 1976, pp. 165–181, illus.

Wihr, Rolf
Restaurieren von Keramik und Glas
Munich: Georg D. W. Callwey, 1977, 275 pp., illus.

Woods, Paul
"The Tank Furnace Built and Used in the Royal College of Art Glass Department"
Presented at *Working with Hot Glass*, international conference organized by the Crafts Advisory Committee and Royal College of Art, London, Sept. 1976.

ANCIENT AND ISLAMIC

Anonymous
"The Antiquities Market; Market Alert: Theft of Archaeological Items from Cyprus"
Journal of Field Archaeology, 3, No. 3, 1976, pp. 335–339, illus.
Stolen beads, pendant from Cyprus.

Alarcão, Jorge de
"Vidros romanos procedentes da Colecção do Rei D. Manuel"
Conimbriga, 15, 1976, pp. 55–61, illus.

Alarcão, Jorge de *et al.*
Fouilles de Conimbriga; VI; Céramiques, diverses et verres
Paris: Diffusion E. de Boccard, 1976, 258 pp., illus.

Alexandre, Olga
"Ayaskaphai etouz 1971"
Archaiologikon Deltion, 27, 1972, pp. 85–146, illus.
Band vessels from an Athenian tomb.

Antique Glass in the Hermitage Collection
Leningrad: Aurora Art Publishers, 1974, folder of 16 postcards with separate text by N. Kunina in Russian and English.

Auth, Susan H.
Ancient Glass at the Newark Museum
Newark, N.J.: The Newark Museum, 1976, 238 pp., illus.

Avi-Yonah, Michael and Stern, Ephraim
Encyclopedia of Archaeological Excavations in the Holy Land, Volume III
Englewood Cliffs, N.J.: Prentice-Hall, Inc., 1977, 300 pp., illus.
Numerous glass vessels, mosaics, etc.

Baatz, Dietwulf
"Das Kastell Munningen im Nördlinger Ries"
Saalburg Jahrbuch, 33, 1976, pp. 11–62, illus.
Roman finds.

Barag, Dan
"The Glass Vessels"
Beth She'arim, Volume III, The Excavations 1953–1958 by Nahman Avigad, Rutgers University Press, New Brunswick, N.J., 1976, pp. 198–209, illus.
Mostly free-blown, green vessels, and several hundred fragments.

Berger, Ludwig
"Découvertes archéologiques récentes de verres: Suisse"
Bulletin de l'Association Internationale pour l'Histoire du Verre, 7, 1973–1976, pp. 183–186.

Berlin. Staatliche Museen, Frühchristlich-byzantinische Sammlung (Bode-Museum)
Byzantinische Kostbarkeiten
Berlin: Staatliche Museen, Frühchristlich-byzantinische Sammlung, 1977, 95 pp., illus.
Several *Nuppenbecher*.

Bernhard, Helmut
"Ein römisches Gräberfeld des 1.–4. Jhs. bei Neustadt/Lachen-Speyerdorf"
Archäologisches Korrespondenzblatt, 3, No. 1, 1973, pp. 59–65, illus.
A beaker and bottle of the late 4th c.

Bichir, Gh.
"Les Sarmates au Bas-Danube"
Dacia, 21, 1977, pp. 167–197, illus.
Late 2nd–3rd c. A.D. bottle.

Black, J. Anderson
"The Craft of King Tut's Jewels"
Craft Horizons, 37, No. 1, Feb. 1977, pp. 20–23, illus.

Blanchard-Lemée, Michèle
Maisons à mosaiques du quartier central de Djemila (Cuicul)
Paris: Editions Ophrys with the National Center for Scientific Research, [1976?], 260 pp., illus.
Roman mosaics from African Mediterranean area.

Bloesch, Hansjörg and Isler, Hans Peter
"Monte Iato: La sesta campagna di scavo"
Sicilia Archeologica, 9, No. 32, Dec. 1976, pp. 9–23, illus.
A 1st c. bowl with wheel-cut grooves.

Bologna. Museo Civico
Pelagio Palagi; Artista e collezionista
Bologna: Grafis, 1976, 462 pp., illus.
Some Roman tableware and balsamari.

Bonfioli, Mara
"Nota sul vetro islamico"
Arte Veneta, 29, 1975, pp. 4–8, illus.

Boston. Museum of Fine Arts
Romans and Barbarians
Boston: Museum of Fine Arts, 1976, 255 pp., illus.

Bracker, Jörgen
"Zur Rekonstruktion und Deutung des Goldglastellers vom Katharinengraben in Köln"
Festschrift für Waldemar Haberey
Ed. Thea Elisabeth Haevernick and Axel von Saldern
Mainz: Philipp von Zabern, 1976, pp. 5–8, illus.

BRISTOL. MUSEUM & ART GALLERY
Ancient Glass; The Bomford Collection of Pre-Roman & Roman Glass on loan to the City of Bristol Museum & Art Gallery
Bristol: Museum & Art Gallery, 1976, 48 pp., illus.

CATLING, H. W.
"Archaeology in Greece, 1975–76"
Archaeological Reports for 1975–76, 22, 1976, pp. 3–33, illus.
Candlestick unguentaria from 2nd–3rd c. graves at Antimachia on Kos.

CHARLESWORTH, DOROTHY
"The Roman Glass"
Chichester Excavations by Alec Down
Chichester: Phillimore & Co. Ltd., 1974, pp. 133–137, illus.
Variety of fragment finds.

CHARLESWORTH, DOROTHY
"Roman Site at Kirkbridge, Cumberland: The Glass Finger-ring"
Transactions of the Cumberland and Westmoreland Antiquarian and Archaeological Society, 75, 1975, p. 88.

CINCINNATI, OHIO. THE TAFT MUSEUM
Best of 50
Cincinnati: Cincinnati Institute of Fine Arts [1977], n.pag., illus.
Roman mosaic inlay.

COONEY, JOHN D.
"An Egyptian Mosaic Glass Panel"
Boston Museum Bulletin, 74, No. 370, 1976, pp. 110–114, illus.

CULICAN, W.
"Sidonian Bottles"
Levant, 7, 1975, pp. 145–150, illus.

DEICHMANN, FRIEDRICH W.
Ravenna: Haupstadt des spätantiken Abendlandes
Band I: Geschichte und Monumente
Band II: Geschichte und Monumente. Kommentar, 1. Teil
Geschichte und Monumente. Kommentar, 2. Teil
Plananhang zu Band II
Band III: Frühchristliche Bauten und Mosaiken von Ravenna
Wiesbaden: Franz Steiner Verlag, 1958–1976, illus.

DIACONU, GHEORGHE and others
"L'ensemble archéologique de Pietroasele"
Dacia, 21, 1977, pp. 199–220, illus.
A 4th c. facet-cut beaker and several 3rd c. fragments.

DOPPELFELD, OTTO
"Der neue Kölner Konchylienbecher"
Festschrift für Waldemar Haberey
Ed. Thea Elisabeth Haevernick and Axel von Saldern
Mainz: Philipp von Zabern, 1976, pp. 23–28, illus.

ELGAVISH, J.
"Chronique archéologique: Shiqmona; deux campagnes de fouilles"
Revue Biblique, 82, No. 4, 1975, pp. 587–591.
Mosaics, bracelets, vases, pâte de verre beads.

ELGAVISH, J.
"Shikmona"
Revue Biblique, 83, No. 2, 1976, pp. 270–272, illus.

ERDMANN, ELISABETH
"Die Glasfunde von Mezad Tamar (Kasr Gehainije) in Israel"
Saalburg Jahrbuch, 34, 1977, pp. 98–146, illus.

ERIM, K. T. AND REYNOLDS, JOYCE
"The Aphrodisias Copy of Diocletian's Edict on Maximum Prices"
Journal of Roman Studies, 63, 1973, pp. 99–110, illus.
Notes on glass entries by Dorothy Charlesworth.

FREYER-SCHAUENBURG, BRIGITTE
"Die Glasfunde aus Pitane (Çandarli)"
Anatolia, 17, 1973, (1975), pp. 141–175, illus.
Core-formed vessels.

FUKAI, SHINJI
Persian Glass
New York: John Weatherhill, Kyoto: Tankosha, 1977, 66 pp. illus.

FÜLEP, FERENC
Roman Cemeteries on the Territory of Pécs (Sopianae)
Budapest: Akadémiai Kiadó, 1977, 64 pp., illus.
Beads, bangles, bottles, jugs, etc.

GERMAIN, SUZANNE
Les Mosaïques de Timgad
Paris: Éditions du Centre National de la Recherche Scientifique, 1973, 170 pp., illus.

GOLDSTEIN, SIDNEY
"Glass Fragments from Tell Hesbân"
Andrews University Seminary Studies, 14, No. 1, Spring 1976, pp. 127–132, illus.

GOLDSTEIN, SIDNEY
"Mosaic Glass in Antiquity"
Glass Paperweight Symposium, June 1976, The John Nelson Bergstrom Art Center and Museum, Neenah, Wisconsin, pp. 10–12.

GOMOLKA, GUDRUN
"Découvertes archéologiques récentes de verres: Bulgarie"
Bulletin de l'Association Internationale pour l'Histoire du Verre, 7, 1973–1976, pp. 167–168.
3rd–7th c. finds at Limeskastell Iatrus.

GOODBURN, R., *et al.*
"Roman Britain in 1975"
Britannia, 7, 1976, pp. 290–392, illus.
Cameo fragment of Augustan date; window glass.

GREENAWAY, J. AND HENIG, M.
"Notes: A Moulded Glass Cameo from Silchester"
Britannia, 6, 1975, pp. 209–210, illus.

GROSE, D. F.
"Découvertes archéologiques récentes de verres: Italie"
Bulletin de l'Association Internationale pour l'Histoire du Verre, 7, 1973–1976, pp. 175–182, illus.
Variety of glass found at Cosa.

GUREVICH, F. D.
"Dea etapa v istorii drevnerusskih gorodov Poneman'ja"
Kratkie Soobshscheniia, No. 146, pp. 25–31, illus.
Islamic beads and vessels.

HAEVERNICK, THEA ELISABETH
"Doppelköpfchen"
Wissenschaftliche Zeitschrift der Universität Rostock, 17, No. 7/8, 1968, pp. 647–653, illus.
English summary.
Amulets with double heads, Carthage 4th c. B.C.

HAEVERNICK, THEA ELISABETH
"Gedanken zur frühesten Glasherstellung in Europa"
Jahrbuch des Römisch-Germanischen Zentralmuseums Mainz, 21, 1974, Festschrift Hundt, Teil 1, pp. 205–209, illus.

HAEVERNICK, THEA ELISABETH AND CALMEYER, PETER
"Ein gläserner Huwawa"
Archaeologische Mitteilungen aus Iran, 9, 1976, pp. 15–19, illus.
Pendants.

HAEVERNICK, THEA ELISABETH AND HOPF, MARIA
"Glas und Bronzeamulett aus Karthago"
Festschrift für Waldemar Haberey
Ed. Thea Elisabeth Haevernick and Axel von Saldern
Mainz: Philipp von Zabern, 1976, pp. 29–32, illus.

HAMBURG. MUSEUM FÜR KUNST UND GEWERBE
Kunst der Antike
Mainz: Philipp von Zabern, 1977, 556 pp., illus.
Includes Roman glass.

HARRISON, EVELYN
"The Portland Vase: Thinking It Over"
Essays in Archaeology and the Humanities
Mainz: Philipp von Zabern, 1976, pp. 131–142, illus.

HATT, J. J. AND THÉVENIN, A.
"Trouvailles de tombes Gallo-Romaines à Koenigshoffen"
Cahiers Alsaciens d'Archéologie, d'Art, et d'Histoire, 12, 1968, pp. 31–38, illus.
Bottles, beads.

HAUPT, DOROTHEA
"Blauer Glaskrater aus Bonn"
Festschrift für Waldemar Haberey
Ed. Thea Elisabeth Haevernick and Axel von Saldern
Mainz: Philipp von Zabern, 1976, pp. 51–57, illus.

HESSEN, OTTO VON
Primo contributo alla archaeologia longobarda in Toscana; le necropoli
Florence: Leo S. Olschki Editore, 18, 1971, 80 pp., illus.
Beads.

HINZ, HERMANN
"Eine gefleckte Glaskanne aus Xanten"
Archaeologie en Historie, ed. by Brunsting, Bussum: Fibula-van Dischoeck, 1973, No. 372, pp. 343–348, illus.

ISHIGURO-KŌJIRŌ
"A Speculation on the Origin of Transparent Glass"
Glass (Garasu), No. 2, Sept. 1976, pp. 11–17, illus., English summary.
Pre-Hellenistic cut glass bowls from Syria and Iran in a Tokyo collection.

JANSSEN, W.
"Jahresbericht 1974: Fränkische Zeit (Merowingerzeit)"
Bonner Jahrbücher, 176, 1976, pp. 424–430, illus.
A cone beaker from Stommeln.

KOCH, ROBERT
"Spätrömische Ösenperlen aus kobaltblauem Glas"
Festschrift für Waldemar Haberey
Ed. Thea Elisabeth Haevernick and Axel von Saldern
Mainz: Philipp von Zabern, 1976, pp. 71–78, illus.

KOLENDO, JERZY
"Parcs à huitres et viviers sur un flacon en verre du Musée National de Varsovie"
Études et Travaux, (Travaux du Centre d'Archéologie Méditerranéenne de l'Académie Polonaise des Sciences), 18, No. 9, 1976, pp. 144–158, illus.
Engraved bottle, 3rd–4th c.

KRITZECK, JAMES
"Two Early Arabic Glass Weights"
Andrews University Seminary Studies, 14, No. 1, 1976, pp. 157–162, illus.

Kul'tura i iskusstvo antichnogo mira
Leningrad: Aurora, 1974, 191 pp., illus.
Ancient works of art in The Hermitage.

LA BAUME, PETER
"Aussergewöhnliche Stücke römischer Kleinkunst aus einer Privatsammlung"
Festschrift für Waldemar Haberey
Ed. Thea Elisabeth Haevernick and Axel von Saldern
Mainz: Philipp von Zabern, 1976, pp. 79–83, illus.
Roman miniatures.

LACROIX, BERNARD
"Mobilier d'un habitat du IVe siècle aux Fontaines-Salées"
Revue Archéologique de l'Est et du Centre-est, 19, 1968, pp. 191–233, illus.
Bottle, goblet fragments.

LAURÉN, SYNNÖVE, FERENIUS, J., AND PERSSON, A.
"Några nya fynd från Stockholm"
Fornvannen, 65, 1970, pp. 108–114, illus. English summary.
Funnel beaker.

LEHECKOVA, EVA
"Découvertes archéologiques récentes de verres: Tchécoslovaquie"
Bulletin de l'Association Internationale pour l'Histoire du Verre, 7, 1973–1976, pp. 187–198, illus.
Finds at many sites from Bronze Age to Slavonic Period.

LÖBER, HANS
"Hatte Plinius d. Ä. recht mit seinem Bericht über das Entstehen des Glases?"
Festschrift für Waldemar Haberey
Ed. Thea Elisabeth Haevernick and Axel von Saldern
Mainz: Philipp von Zabern, 1976, pp. 85–88, illus.
Pliny's account of the origin of glass.

MAIOLI, MARIA
"I vetri della necropoli romana de 'Le Palazzette' nel territorio di Classe"
Felix Ravenna, Series 4, 7/8, 1974, pp. 15–36, illus.
Large group of balsamari.

MALIBU, CALIFORNIA.
THE J. PAUL GETTY MUSEUM
Catalogue of the Ancient Art in the J. Paul Getty Museum (text by Cornelius Vermeule and Norman Neuerburg)
Malibu: The J. Paul Getty Museum, 1973, 60 pp., illus.
Mosaics.

MARIEN, M. E.
"Objets de bronze comme 'antiquités' dans les mobiliers de tumuli belgo-romains"
Bulletin des Musées Royaux d'Art et d'Histoire, 46, No. 6, 1974, pp. 9–19, illus.
Includes glass vessels.

MENTASTI, ROSA BAROVIER
"Riflessi d'arte antica nei vetri ottocenteschi di una collezione privata veneziana"
Giornale Economico della Camera di Commercio Industria Artigianto, Agricoltura di Venezia, No. 1, June/July 1977, pp. 43–56, illus.
Salviati collection.

Meunier, Lutz
"Berlin/Antikenmuseum; Antike Gläser"
Die Weltkunst, 47, No. 2, Jan. 1977, p. 81, illus.
Review of an exhibit.

Muncie, Indiana. Ball State University Art Gallery
Ancient Glass Collections (text by Clara Wiltse)
Muncie, Ind.: Ball State University, 1977, n.pag., illus.

Parlasca, Klaus
"Zur Verbreitung ptolemäischer Fayencekeramik ausserhalb Ägyptens"
Jahrbuch des Deutschen Archäologischen Instituts, 91, 1976 (1977), pp. 135–156, illus.
Glass and faience from Canosa.

Paškvalin, Veljko
"Tombeaux antiques tardifs de Jajce"
Wissenschaftliche Mitteilungen des Bosnisch-Herzegowinischen Landesmuseums, 5, A, 1975, pp. 69–79, illus.
Roman head flask and bottle with spout.

Perin, Patrick
"Découvertes archéologiques récentes de verres: France"
Bulletin de l'Association Internationale pour l'Histoire du Verre, 7, 1973–1976, pp. 169–174, illus.
Late Roman finds at Omont.

Pfeffer, Waltraud v.
"Ein Konchylienbecher in Schlangenfadentechnik aus Worms"
Festschrift für Waldemar Haberey
Ed. Thea Elisabeth Haevernick and Axel von Saldern
Mainz: Philipp von Zabern, 1976, pp. 95–99, illus.
Footed beaker in snake-threaded style.

Pirling, Renate
"Klothos Kunkel"
Festschrift für Waldemar Haberey
Ed. Thea Elisabeth Haevernick and Axel von Saldern
Mainz: Philipp von Zabern, 1976, pp. 101–109, illus.
Beads and flasks, Krefeld-Gellep.

Plesničar-Gec, Ljudmila
Severno Emonsko Grobišče (The Northern Necropolis of Emona)
Ljubljana: Narodni Muzej v Ljubljani, 1972, 272 pp., illus.

Rahmani, L. Y.
"A Phoenician Scarab from Ashkelon"
'Atiqot, Vol. 11, 1976, pp. 110–111, illus.

Rahmani, L. Y.
"Roman Tombs in Nahal Raqafot, Jerusalem"
'Atiqot, Vol. 11, 1976, pp. 77–88, illus.
Glass bracelet and some 3rd c. fragments.

Rau, H. Günter
"Konische Glasbecher mit schrägen Wandrillen als Beleg frühfränkischer Glasproduktion"
Festschrift für Waldemar Haberey
Ed. Thea Elisabeth Haevernick and Axel von Saldern
Mainz: Philipp von Zabern, 1976, pp. 111–120, illus.
Beaker with drilled diagonal grooves.

Rau, H. Günter
"Spätantike Glasfunde im Karpatenraum"
Zeitschrift für Ostforschung, 24, No. 3, 1975, pp. 464–485, illus.
4th c. Roman and barbarian glass in Carpathian area.

Rounitch, A. P.
"Zakhoronenie vozhdia epokhi rannego srednevekovyia iz Kislovodskoj kotloviny" (La sepulture d'un chef de la phase initiale du Moyen Âge dans le bassin de Kislovodsk)
Sovetskaia Arkheologiia, No. 3, 1976, pp. 256–266, illus., French summary.
Vessels and beads of the 4th–5th c.

Saldern, Alex von
"Glasrhyta"
Festschrift für Waldemar Haberey
Ed. Thea Elisabeth Haevernick and Axel von Saldern
Mainz: Philipp von Zabern, 1976, pp. 121–126, illus.

Schieb, Gustl
"Aus der Geschichte des Glases"
Der Glaserlehrling, 28, No. 11, 1975, pp. 1–2.

Schindler, Reinhard
Führer durch das Landesmuseum Trier
Trier: Rheinisches Landesmuseum Trier, 1977, 112 pp., illus.

Schmidt, Berthold
"Découvertes archéologiques récentes de verres: Allemagne (R.D.A.)"
Bulletin de l'Association Internationale pour l'Histoire du Verre, 7, 1973–1976, pp. 153–157, illus.

Seefried, Monique
"Les pendentifs en verre façonnés sur noyau du Musée National du Bardo et du Musée National de Carthage"
Karthago, 17, 1976, pp. 37–66, illus.

Sorokina, N. P.
"Ein Glaspokal mit Iphigeniens Namen aus Pantikapaion"
Festschrift für Waldemar Haberey
Ed. Thea Elisabeth Haevernick and Axel von Saldern
Mainz: Philipp von Zabern, 1976, pp. 145–152, illus.

Sorokina, N. P.
"Stekliannye perstni iz Prikuban'ia v sobranii GIM"
Archeologicheskii Sbornik, 40, Sovetskaia Rossiia, 1966, pp. 71–73, illus.
Persian ring.

Sussman, Varda
"A Burial Cave at Kefar 'Ara"
'Atiqot, Vol. 11, 1976, pp. 92–101, illus.
Vessels of the 2nd–6th c.

Symonovich, E. A.
"Stekliannaia posuda iz podneprovsko prichernomorskikh pamiatnikov cherniakhovskoĭ kul'tury" (La verrerie des monuments de la culture de Tcherniakhovo, dans les regions du Dniepr et de la mer Noire)
Sovetskaia Arkheologiia, 1, 1977, pp. 176–186, illus., French summary.

Tokyo. Gallery Mikazuki
Exhibition of Ancient Glass and Glazed Ware Collected by G. McKinley and K. Ishiguro
Tokyo: Gallery Mikazuki, 1975, n.pag., illus.

Tomasevic, Teodora
"Ein Glasschmelzofen in den Äusseren Reben, Kaiseraugst AG"
Regio Basiliensis, 18, No. 1, 1977, pp. 243–252, illus.
4th c. glassmelting furnace in Rhine area.

Tomobe-Tadashi
"Mycenaean Glass Beads"
Glass (Garasu), No. 2, Sept. 1976, pp. 18–21, illus., English summary.

University Park, Pa. Museum of Art
Ancient Glass—Classical, Sassanian, and Islamic from the Collection of Jerome Strauss
University Park, Pa.: The Pennsylvania State University Museum of Art, 1977, n.pag., illus.

VAN DEN HURK, L. J. A. M.
"The Tumuli from the Roman Period of Esch, Province of North Brabant, I"
Berichten van de Rijksdienst voor het Oudheidkundig Bodemonderzoek, 23, 1973 (1975), pp. 189–236, illus.
Roman square bottles.

VORONOV, N.
"Gienos"
Sovetskaia Arkheologiia, 4, 1976, pp. 42–55, illus.
A 4th–5th c. blob-decorated cup.

WEINBERG, S.
"Tel Anafa"
Revue Biblique, 77, 1970, pp. 381–383, illus.

WHITE, DONALD
"Excavations in the Sanctuary of Demeter and Persephone at Cyrene. Fourth Preliminary Report"
American Journal of Archaeology, 80, No. 2, Spring 1976, pp. 165–181, illus.
Ram's head pendant.

WHITEHOUSE, HELEN
The Dal Pozzo Copies of the Palestrina Mosaic
Oxford: B.A.R. Supplementary Series 12, 1976, 97 pp., illus.

WISEMAN, JAMES AND MANO-ZISSI, DJORDJE
"Stobi: A City of Ancient Macedonia"
Journal of Field Archaeology, 3, No. 3, 1976, pp. 269–302, illus.
Beads.

YOSHIMIZU, TSUNEO
"Type Classifications of Roman Glass of Germania"
Glass (Garasu), Association for Glass Art Studies, No. 1, Dec. 1975, pp. 9–16, illus., English summary.

MEDIEVAL

Akten des 10. Internationalen Colloquiums des Corpus Vitrearum Medii Aevi
Stuttgart: Corpus Vitrearum Medii Aevi, 1977, 71 pp.
Resumés of papers presented on medieval and stained glass.

ALS HANSEN, BIRGIT
"Middelalderlige glasmalerier"
Hikuin, 1, 1974, pp. 87–96, illus.
Examples and locations of Danish painted glass.

BESTEMAN, J. C.
"Carolingian Medemblik"
Berichten van de Rijksdienst voor het Oudheidkundig Bodemonderzoek, 24, 1974, (1976), pp. 43–106, illus.
9th c. beakers and beads.

BRONICKA-RAUHUT, JADWIGA AND RAUHUT, LECHOSLAW
"Wczesnósredniowieczne cmentarzysko szkieletowe w Starogrodzie woj. Siedlce" (An early Medieval cemetery at Starogród Siedlce Voivodship)
Wiadomósci Archeologiczne, 42, No. 1, 1977, pp. 58–86, illus., English summary.
12th–13th c. beads.

CALLMER, JOHAN
Trade Beads and Bead Trade in Scandinavia ca. 800–1000 A.D.
Bonn: Rudolf Habelt, 1977, 229 pp., illus.

FOY, DANIÈLE
"L'artisanat du verre creux en Provence médiévale"
Archéologie Médiévale, 5, 1975, pp. 103–138, illus.

GRIESS, JULES
"Le cimetière mérovingien d'Avolsheim"
Cahiers Alsaciens d'Archeologie, d'Art, et d'Histoire, 12, 1968, pp. 43–54, illus.
Beads of pâte de verre.

HAN, VERENA
"Die Glaserzeugung des 14. und 15. Jh. in Dubrovnik"
La formation et le dévéloppement des metiers au Moyen âge/Ve–XIVe siècles/ .Akadémiai Kiadó. Budapest, 1977, pp. 127–134.

HILGER, HANS PETER
"Ein Reliquienkreuz aus Bergkristall in St. Gereon in Köln"
Festschrift für Waldemar Haberey
Ed. Thea Elisabeth Haevernick and Axel von Saldern
Mainz: Philipp von Zabern, 1976, pp. 59–64, illus.
Rock crystal reliquary, 13th c.

HOLDSWORTH, PHILIP
"Saxon Southampton; a New Review"
Medieval Archaeology, 20, 1976, pp. 26–61, illus.
Rim fragments including some that are unusual in form or decoration. Late 7th–early 10th c.

LÁSZLÓ, GYULA
The Art of the Migration Period
Coral Gables, Florida: University of Miami Press, 1974, 135 pp., illus.
Glass inlays and beads.

LUNDSTRÖM, AGNETA
"Bead Making in Scandinavia in the Early Middle Ages"
Kungl. Vitterhets, Historie och Antikvitets Akademien, 61, n.d. [1976], pp. 3–19, illus.

LUNDSTRÖM, PER
"Paviken I bei Västergarn—Hafen, Handelsplatz und Werft"
Vor- und Frühformen der europäischen Stadt im Mittelalter (Abhandlungen der Akademie der Wissenschaften in Göttingen), II, Symposium in Reinhausen, 1972, pp. 82–93, illus.
Beads from Gotland.

NEUFFER-MÜLLER, CHRISTIANE
"Fränkische Rüsselbecher aus Württemberg"
Festschrift für Waldemar Haberey
Ed. Thea Elisabeth Haevernick and Axel von Saldern
Mainz: Philipp von Zabern, 1976, pp. 89–94, illus.

NEW YORK. THE METROPOLITAN MUSEUM OF ART
Age of Spirituality
New York: The Metropolitan Museum of Art, 1977, 96 pp., illus.
Mosaics, enamels, gold-glasses of 3rd–7th c.

OLMEROVÁ, HELENA
"A Find of Mediaeval Glass in the Old Town of Prague"
Glass Review, 32, No. 5, 1977, pp. 24–27, illus.

PERIN, PATRICK
"Quelques objets exceptionnels provenant des tombes de chefs du cimetière mérovingien de Mézières"
Revue Historique Ardennaise, 4, July–Dec. 1970, pp. 71–77, illus.
5th–6th c. drinking horn with white trailed decoration.

PERIN, PATRICK, and others
"Ensembles archéologiques mérovingiens de la région Ardennaise: Les sites de Villers-Semeuse, Aubigny-les-Pothées, Jandun, Fagnon et Semide"
Revue Historique Ardennaise, 5, Jan.–June 1971, pp. 19–61, illus.
Footless beakers and bowls.

PLETNEVA, S. A. AND NIKOLAIENKO, A. G.
"Volokonovskii drevnebolgarskii mogil'nik" (L'ancien champ d'inhumation bolgare de Volokonovo)
Sovetskaia Arkheologiia, No. 3, 1976, pp. 279–298, illus., French summary.
Late 9th–10th c. beads.

POPA, RADU
"Streisingeorgiu. Ein Zeugnis rumänischer Geschichte des 11.–14. Jahrhunderts im Süden Transsilvaniens"
Dacia, 20, 1976, pp. 37–61, illus.
A series of glass earrings.

RAU, HERMANN GÜNTER
"Alamannische Grabfunde im Bachgau, Landkreis Aschaffenburg"
Archäologisches Korrespondenzblatt, 3, No. 2, 1973, pp. 237–242, illus.
6th and 7th c. beads and a conical beaker.

SCHIEB, GUSTL
"Glas im Mittelalter"
Der Glaserlehrling, 28, No. 12, 1975, pp. 2–3.

SCHÜTTE, SVEN
"Mittelalterliches Glas aus Göttingen"
Zeitschrift für Archäologie des Mittelalters, 4, 1976, pp. 101–117, illus., English summary.
15th c. vessels and fragments.

SIX, HERBERT
"Spätmittelalterliche Glashütten im Hils bei Grünenplan mit Farbglasproduktion"
Festschrift für Waldemar Haberey
Ed. Thea Elisabeth Haevernick and Axel von Saldern
Mainz: Philipp von Zabern, 1976, pp. 129–144, illus.

STEPHAN, HANS-GEORG
"Découvertes archéologiques récentes de verres: Allemagne (R.F.A.)"
Bulletin de l'Association Internationale pour l'Histoire du Verre, 7, 1973–1976, pp. 158–166, illus.
Glass finds at Höxter/Weser.

VERWERS, W. J. H.
"A Merovingian Cemetery in Veldhoven, Province of North Brabant"
Berichten van de Rijksdienst voor het Oudheidkundig Bodemonderzoek, 23, 1973 (1975), pp. 313–335, illus.
Beaker, beads.

VOSE, RUTH HURST
"Glassmaking at Kingswood, Delamere, Cheshire"
Vale Royal Abbey and House, Winsford Local History Society in conjunction with the Michaelmas Trust, Summer 1977, pp. 34–36. 13th–15th c. clear window and stained glass.

YPEY, J.
"Das fränkische Gräberfeld zu Rhenen, Prov. Utrecht"
Berichten van de Rijksdienst voor het Oudheidkundig Bodemonderzoek, 23, 1973 (1975), pp. 289–312, illus.
Beakers, beads.

RENAISSANCE TO PRESENT

GENERAL

ANONYMOUS
"Christmas Tree History; Antique Christmas Ornaments"
The Collector, 9, No. 321, Dec. 1977, p. 1, illus.
"Collectors' Questions"
Country Life, 161, No. 4163, April 14, 1977, p. 926, illus.
Glass bust of Henri, Comte de Chambord, by French sculptor, Jean Louis Veray.
"Frost Trade Bead Exhibit"
The Bead Journal, 3, No. 2, 1977, p. 31, illus.
Collection at Illinois State Museum.
"Spotlight on Paperweights"
The Acorn, 11, No. 2, May 1977, p. 1, illus.

ANDERSEN, DANNESBOE
Gammelt Dansk Apoteksinventar
Copenhagen: Ejnar Munksgaard, 1944, 416 pp., illus.
Apothecary glassware, 17th–19th c.

BAMFORD, JOAN
"Antique Paste Jewelry"
Spinning Wheel, 33, No. 9, Nov. 1977, pp. 8–13, illus.

BÁRTA, JAN
"Les rapports de la verrerie de Bohême avec celle de Venise au XVIIIe siècle"
Acta Historiae Rerum Naturalium Necnon Technicarum (Prague), No. 6, 1972, pp. 55–63.

BUCHHEIM, LOTHAR-GÜNTHER
"Tausend Träume unter Glas"
Die Zeit, 26, 1974, pp. 14–16, illus.
Paperweights.

BUTIKOV, GEORGE
St. Isaac's Cathedral, Leningrad
Leningrad: Aurora Art Publishers, 1974, 143 pp., illus.
19th c. Russian and Italian mosaics.

CIEPIELA, SŁAWOMIRA
Szkło Osiemnastowieczne Starej Warszawy
Warsaw: Państwowe Wydawnictwo Naukowe, 1977, 141 pp., illus., English summary.
18th c. glass of Old Warsaw.

CLELAND, CHARLES, ed.
An Historic Burial Locality in Mackinac County, Michigan
East Lansing, Mich.: Publications of the Museum, Michigan State University, 1971, 147 pp., illus.
Beads.

DAVIS, FRANK
"Modern Tribute to an Ancient Game" (Talking About Salerooms series)
Country Life, 161, No. 4161, March 31, 1977, pp. 778–779, illus.
German and *façon de venise* items auctioned.

DRESDEN. MUSEUM FÜR KUNSTHANDWERK
Kunsthandwerk und Industrieform des 19. und 20. Jahrhunderts
Dresden: Staatliche Kunstsammlung, n.d., 258 pp., illus.
Glass, pp. 71–94, by Gisela Haase.

ENGLE, ANITA
Readings in Glass History No. 8
Jerusalem: Phoenix Publications, 1977, 87 pp., illus.
Continental glassmakers who found refuge in England during the Elizabethan period: Jean Carré, Verzelini, etc.

FENSTERMAKER, G. B.
Mardi Gras Colored Bead Chart
Lancaster, Pa.: G. B. Fenstermaker, 1977, n.p., illus.

FRACE, CLEONE
"Glass Christmas Tree Ornaments"
The Federation Letter, 7, No. 11, Dec. 1977, pp. 4–6, illus.

GARLAKE, PETER
"An Investigation of Manekweni, Mozambique: Beads in Glass, Shell and Copper"
Azania, 11, 1976, pp. 25–47, illus.

GREENBLATT, THERESA AND NESBITT, DON
"Paperweights at Smith College Art Museum"
Annual Bulletin of the Paperweight Collectors' Association, 1977, pp. 3–10, illus.

GUSTAFSON, ELEANOR
"Museum Accessions"
Antiques, 111, No. 3, March 1977, pp. 486–492, illus.
English-Irish candelabra, Sandwich craquelé pitcher, Clichy millefiori vases, Mt. Washington vases.

HAKS, FRANS
"Signalement van het Nationaal Glasmuseum in Leerdam"
Museum Journaal, 19, No. 3, June 1974, pp. 101–108, illus.
Description of the museum.

HAN, VERENA
"Trgovina Dubrovačkim staklom u Albaniji u XVI veku" (Le commerce du verre de Dubrovnik en Albanie XVIe siècle)
Balcanica, 7, 1976, (Godišnjaka Balkanološkog instituta, Beograd), pp. 269–290, French summary.

HAN, VERENA AND ZECCHIN, LUIGI
"Presenze Balcaniche a Murano e presenze Muranesi nei Balcani"
Balcanica, 6, (Godišnjaka Balkanološkog instituta, Beograd), 1975, pp. 77–97, Serbo-Croatian summary.

HARRIS, J. R.
"Saint-Gobain and Ravenhead"
Great Britain and Her World, Manchester University Press, 1975, Essays in honor of W. O. Henderson, ed. by B. Ratcliffe, pp. 27–70.
A comparative study of French and British plate glass manufactories in the 18th and 19th c.

HAWLEY, JOHN
"Oceans Apart—Yet Alike"
Annual Bulletin of the Paperweight Collectors' Association, 1977, pp. 27–28, illus.
A comparison of French and American weights.

HEACOCK, WILLIAM
1000 Toothpick Holders; A Collector's Guide
Marietta, Ohio: Antique Publications, 1977, 112 pp., illus.

HEINZL, BRIGITTE
"Die Glassammlung der kunsthistorischen Abteilung des OÖ. Landesmuseums"
Jahrbuch des Oberösterreichischen Musealvereines, Linz, 120, No. 1, 1975, pp. 211–224, illus.
Glass collection at Austrian national museum.

HENNIG, WOLFGANG
"Glas—Historismus und die Historismen um 1900"
Silikattechnik, 28, No. 7, July 1977, pp. 220–222, illus.

HERSH, J. JOSEPH
"The Glass Fly Trap"
The Antiques Journal, 32, No. 8, Aug. 1977, p. 21, illus.

HOLLISTER, PAUL
"Confessions of a Collector"
Collector Editions, 5, No. 4, Fall 1977, pp. 31–33, illus.
Paperweights.

INDIANAPOLIS. INDIANAPOLIS MUSEUM OF ART
The Collection of Mr. and Mrs. Herman C. Krannert
Indianapolis: Indianopolis Museum of Art, 1976, n.pag., illus.
Snuff bottles and paperweights.

INGOLD, GERARD
"Homage to Madame Colette"
Annual Bulletin of the Paperweight Collectors' Association, 1977, pp. 11–16, illus.

JOHNSON, MARY ELLEN AND CHAPMAN, BERNADINE
"Old Milk Glass Christmas Tree Lights are Enchanting"
Collectors News, 18, No. 8, Dec. 1977, p. 1+, illus.
U.S. and Japanese manufacture.

LE CLERC, P.
"Development of the French Glass Industry — with Particular Reference to the U.K."
Glass, 53, No. 2, 1976, pp. 46–53.

LOS ANGELES. LOS ANGELES COUNTY MUSEUM OF ART
The Art of Mosaics; Selections from the Gilbert Collection (Text by Alvar Gonzalez-Palacios)
Los Angeles: Los Angeles County Museum of Art, 1977, 143 pp., illus.

MCCAWLEY, PATRICIA
Glass Paperweights
London: Charles Letts & Co. Ltd., 1975, 71 pp., illus.

MCMILLEN, DOROTHY
"The Gentle Glow of Long Ago"
American Collector, 8, No. 9, Sept. 1977, pp. 18–19+, illus.
Victorian fairy lamps.

MCNALLY, PAUL
"Table Glass from the Wreck of the Machault"
Canadian Historic Sites, 16, National Historic Parks and Sites Branch, Parks Canada, Indian and Northern Affairs, Ottawa, 1977, pp. 35–44, illus.
18th c. English and French tablewares.

MANLEY, C. C.
"Experimental Pieces of Victorian Glass"
The Antiques Journal, 32, No. 2, Feb. 1977, pp. 12–15, illus.

MANNONI, EDITH
Fixes et peintures sous verre
Paris: Éditions Ch. Massin, n.d., 64 pp., illus.

MANNONI, EDITH
Opalines
Paris: Éditions Ch. Massin, n.d., 63 pp., illus.

MANNONI, EDITH
Les sulfures et boules presse-papiers
Paris: Éditions Ch. Massin, n.d., 63 pp., illus.

The Masterpieces of the Centennial International Exhibition; Volume II. Industrial Art by Walter Smith
Philadelphia: Gebbie and Barrie, 1876, 521 pp., illus.
Cut, colored and ornamental glass, all European; stained glass.

MILLER, ROBERT W., ed.
Wallace-Homestead Price Guide to Antiques and Pattern Glass
Des Moines, Iowa: Wallace-Homestead Book Co., 1977, 408 pp., illus.

MUCHA, MIRIAM
"The George H. Lorimer Collection of American and European Glass at the Philadelphia Museum of Art"
The Glass Club Bulletin, 121, Dec. 1977, pp. 3–10, illus.

Noël Hume, Ivor
The Wells of Williamsburg: Colonial Time Capsules
Williamsburg, Va.: The Colonial Williamsburg Foundation, 1973, 48 pp., illus.
French, English, and American wine bottles, wineglasses, beer bottles dating from late 18th c.–mid-19th c.

Richards, William
A Day in the New York Crystal Palace
New York: G. P. Putnam & Co., 1853, 168 pp., illus.
Guide to the 1853 exhibition. Glass from many countries.

Sacramento, California.
E. B. Crocker Art Gallery
Tradition and Change in Yoruba Art (text by J. J. Arneson)
Sacramento: E. B. Crocker Art Gallery, 1974, 79 pp., illus.
Beads.

St. Petersburg, Florida.
Museum of Fine Arts
The Art of European Glass from 1600 through 1800
St. Petersburg: Museum of Fine Arts, 1977, n.pag., illus.

Schlumberger, Eveline
"Art et artifice des opalines"
Connaissance des Arts, 306, Aug. 1977, pp. 26–33, illus.
Saint-Louis opalines.

Smith, Allan and Helen B.
Individual Open Salts Illustrated; 1976 Annual
Litchfield, Maine: The Country House, 1976, n.pag., illus.

Smith, Allan and Helen B.
Individual Open Salts Illustrated; 1977 Annual
Litchfield, Maine: The Country House, 1977, n.pag., illus.

Smith, Allan and Helen B.
The Third Book of Individual Open Salts Illustrated
Litchfield, Maine: The Country House, 1976, n.pag., illus.

Smith, Marvin T.
"The Chevron Trade Bead in North America"
The Bead Journal, 3, No. 2, 1977, pp. 15–17, illus.

Spanish Glass in the Hermitage (text by O. Mikhailova)
Leningrad: Aurora Art Publishers, n.pag., illus.

Stratton, Deborah
Mugs and Tankards; A Collector's Guide
London: Souvenir Press, 1975, 144 pp., illus.

Tansley, June
"Victorian Christmas Lights"
Old Bottle Magazine, 10, No. 12, Dec. 1977, pp. 34–35, illus.

Turan, Serafettin
"Venedik' te Türk Ticaret merkezi: Fondaco die Turchi"
Belleten Tuerk Tarih Kurumu, Ankara, 32, 1968, pp. 247–283, illus.
Glass trade between Venice and the Ottoman Empire.

Wade, John
"Some Early Australian Glass"
The Australasian Antique Collector, 17, 1976, pp. 104–106, illus.

Warren, Phelps
"Glass Relating to William III"
Antiques, 112, No. 4, Oct. 1977, pp. 742–749, illus.
Continental and Anglo-Irish glass.

Washington, D. C.
Smithsonian Institution
Traveling Exhibition Service
Naive Art in Yugoslavia
Zagreb: Galerija Primitivne Umjetnosti, n.d., [1977?], n.pag., illus.
Paintings on glass.

ENGLISH–IRISH

Bird, Anthony
Paxton's Palace
London: Cassell & Co., Ltd., 1976, 179 pp., illus.

Boydell, Mary
"The Friendly Brothers' Glass"
Country Life, 161, No. 4170, June 2, 1977, pp. 1539–1540, illus.
Engraved sets made for fraternal Irish societies in late 18th–mid 19th c.

Chance, James Frederick
A History of the Firm of Chance Brothers & Co., Glass and Alkali Manufacturers
London: Spottiswoode, Ballantyne & Co. Ltd., 1919, 302 pp., illus.
19th c. Birmingham firm, sheet and optical glass.

The Crystal Palace
London: T. Nelson and Sons, Nelsons' Pictorial Guide-Books for Tourists, n.d. [1867?], 46 pp., illus.
The Sydenham Crystal Palace.

Davis, Frank
"A Marriage of Japanese and Italian" (Talking About Salerooms Series)
Country Life, 161, No. 4159, March 17, 1977, pp. 638–639, illus.
An 18th c. inscribed marriage decanter.

"Marvels in Blue and White" (Talking About Salerooms Series)
Country Life, 162, No. 4184, Sept. 8, 1977, pp. 612–613, illus.
Late 18th c. cabinets mounted with 42 reverse paintings on glass.

"Painter with an Intimate Touch" (Talking About Salerooms Series)
Country Life, 161, No. 4154, Feb. 10, 1977, pp. 318–319, illus.
Beilby goblet.

"The Transforming Brush of Terborch" (Talking About Salerooms Series)
Country Life, 162, No. 4177, July 21, 1977, pp. 138–139, illus.
Auction of a cameo plaque carved by George Woodall.

English Drinking Glasses in the Ashmolean Museum, Oxford
Oxford: Ashmolean Museum, 1977, 32 pp., illus.

Fitzpatrick, Paul
"Waterford Glass"
The Antiques Journal, 32, No. 5, May 1977, pp. 16–19+, illus.

Fitzpatrick, Paul
"Waterford Glass House; Part 2"
The Antiques Journal, 32, No. 6, June 1977, pp. 30–32+, illus.

Josephs, Zoë
"Jewish Glassmakers"
The Jewish Historical Society of England, 25, 1977, pp. 107–119, illus.
Bristol and Birmingham families.

Lacher, Emanuel
"Paperweight Detective Story"
Annual Bulletin of the Paperweight Collectors' Association, 1977, pp. 29–30, illus.
English paperweight with portrait of Duke of Wellington.

London. The Royal Academy of Arts. Diploma Galleries.
Victorian and Edwardian Decorative Art: The Handley-Read Collection
London: The Royal Academy of Arts, 1972, 139 pp., illus.
19th c. Stourbridge glass.

London. 37 Bury Street, St. James's.
James Glass; China & Enamel Painter 1718–1780
London: Albert Amor Ltd., 1977, 45 pp., illus.
Decorated porcelain and glass.

Manchester. Athenaeum Museum of Ceramics
Royal Commemorative Pottery and Glass
Manchester: City of Manchester Cultural Services, 1977, n.pag., illus.

Manley, C. C.
"Acid-Etched Glass: An Underrated Art"
The Antiques Journal, 32, No. 9, Sept. 1977, pp. 20–22+, illus.
Stevens and Williams Co.
"Alabaster Art Glass"
The Antiques Journal, 32, No. 6, June 1977, pp. 22–23, illus.
Stevens and Williams Co.
"Artists and Glass"
The Antiques Journal, 32, No. 12, Dec. 1977, pp. 28–30, illus.
Enameling and painting at Stevens and Williams Co. 1880's–1930's.
"English Epergnes"
The Antiques Journal, 32, No. 8, Aug. 1977, pp. 14–17, illus.
"Stevens and Williams' Applied Decorations on Art Glass"
The Antiques Journal, 32, No. 5, May 1977, pp. 27–29, illus.

Phipps, Jennifer
"The G. Gordon Russell Glass Collection"
Art Bulletin of Victoria, 1973–74, pp. 35–41, illus.
Mostly British, 17th–18th c., collection in National Gallery of Victoria, Melbourne.

Pickford, Ian
"Claret Jugs through the Ages"
Antique Dealer and Collectors Guide, Nov. 1977, pp. 95–100, illus.

Rakow, Dr. and Mrs. Leonard
"James and William Tassie: 18th Century Glass Jewelers and Portraitists"
Glass Paperweight Symposium, June 1976, The John Nelson Bergstrom Art Center and Museum, Neenah, Wisconsin, pp. 25–41.

Stuart, W. E. C.
"Crystal Makers in the Village of Wordsley"
Glass Technology, 18, No. 2, April 1977, pp. 28–34, illus.

Truro, Cornwall. The County Museum.
The English Glass Bottle through the Ages
Truro: The County Museum, 1976, 50 pp., illus.

Webb, Graham
"Glass Novelties from the 19th Century"
The Antique Dealer and Collectors Guide, Dec. 1977, pp. 118–121, illus.

GERMAN–BOHEMIAN–AUSTRIAN–CZECHOSLOVAKIAN

Allen, Francis, comp.
Schebek's Böhmens Glasindustrie und Glashandel (1878): *An Index of Personalities, Glassworks and Companies*
Bellemead, Md.: Francis N. Allen, 1977, n.pag.
An index to a base source of early Bohemian glassworks and workers.

Bletter, Rosemarie Haag
"Paul Scheerbart's Architectural Fantasies"
Journal of the Society of Architectural Historians, 34, No. 2, May 1975, pp. 83–97, illus.
German Expressionist glass architect.

Bloss, Otto
Die älteren Glashütten in Südniedersachsen
Hildesheim: Verlag August Lax, 1977, 201 pp., illus.
15th–18th c. history of old Saxony glasshouses.

Braunschweig. Herzog Anton Ulrich-Museum
Kunst des Barock in Polen aus dem Nationalmuseum in Warschau
Braunschweig (BRD): Herzog Anton Ulrich-Museum, 1974, 43 pp., illus.
18th c. glassware from the Warsaw National Museum.

Broul, Julius
"Sklářský průmysl na Teplicku"
Sklář a Keramik, 27, 1977, No. 1, pp. 7–12, illus.
Glass industry in the region of Teplice.

Brožová, Jarmila
"Friedrich Egermann (1777–1864)"
Umění a Řemesla, 1, 1977, pp. 27–32, illus.
Bohemian glassmaker.

Brožová, Jarmila
"Friedrich Egermann and the Glass of His Time"
Glass Review, 32, No. 6, 1977, pp. 10–12, illus.

Chojnacka, Halina and Chrzanowska, Paulina
"Sächsisch beeinflusstes Barockglas in Polen"
Dresdener Kunstblatter, 18, No. 3, 1974, pp. 78–83, illus.

Davis, Frank
"Oriental Whimsies" (Talking About Salerooms Series)
Country Life, 161, No. 4173, June 23, 1977, pp. 1726–1727, illus.
Bohemian covered bowl, decorated by Preissler.

Distelberger, Rudolf
"Die Sarachi-Werkstatt und Annibale Fontana"
Jahrbuch der Kunsthistorischen Sammlungen in Wien, 71, 1975, pp. 95–164, illus.
Late 16th c. engraved rock crystal.

Dreier, Franz-Adrian
"Fünf Silbermontierte Prunkvasen aus Goldrubinglas"
Berliner Museen, 19, 1969, pp. 78–83, illus.
17th c. German.

Dreier, Franz-Adrian
"Kunstgewerbemuseum — Neuerwerbungen und Schenkungen 1972/73"
Jahrbuch Preussischer Kulturbesitz, 11, 1973, pp. 267–276, illus.
Reverse painting on glass by Sprüngli; pokals, beakers, etc.

Dresden. Staatliche Kunstsammlungen Dresden
Das Grüne Gewölbe
Dresden: Staatliche Kunstsammlung, 1976, 121 pp., illus.
18th c. engraved pokal and beakers in Saxon treasury.

Eckert, Christian
Gedenkblätter zum fünfzigjährigen Jubiläum der Aktiengesellschaft
Cologne: Kölner Verlags-Anstalt und Druckerei A.G., 1922, 67 pp., illus.
Anniversary booklet: Rheinische Glashütten-Aktiengesellschaft Köln-Ehrenfeld.

Essen. Villa Hügel
Kunst des Barock in Böhmen
Recklinghausen: Aurel Bongers, 1977, 316 pp., illus.
Includes work by Ignaz Preissler.

Glass and Pottery in Czechoslovakia (text by Jarmila Brožová)
Prague: Uniprojekt, n.d., [1977], n.pag., illus.

Haase, Gisela
"Das Inventar der Willkommen- und Gesundheitsgläser von Schloss Moritzburg und die Ausstellung 'Sachsisches Glas' im Schloss Pillnitz"
Dresdener Kunstblätter, 18, No. 3, 1974, pp. 66–74, illus.
17th & 18th c. Saxon.

Han, Verena
"'Pjatogrla Stakla' iz gornjeg grada beogradske tvrćave"
Arkheološki Muzej na Makedonija, Zbornik posveten na dimče koco, 6–7, 1975, pp. 317–326, illus., French summary.
Kuttrolf-type bottles, early 18th c., found in Belgrade.

Hoffman, Rudolf
Das Museum für Glaskunst Lauscha; Ein Spiegelbild volkskünstlerischen Schaffens
Lauscha: Museum für Glaskunst, 1971, 48 pp., illus.

Hoffman, Rudolf
Zur sozialen Lage der Werktätigen in der Lauschaer Glasindustrie unter den Bedingungen kapitalistischer Produktionsverhältnisse
Lauscha: Museum für Glaskunst, 1977, 100 pp., illus.

Jablonec nad Nisou. Museum Skla a Bižuterie.
Bedřich Egermann 1777–1864
Jablonec nad Nisou: Museum Skla a Bižuterie, n.d. [1977], n.pag., illus.
Bohemian glassmaker.

Jablonec nad Nisou. Museum Skla a Bižuterie.
Vladimír Linka; sklářský výtvarník, glyptik, sochař, pedagog 1912–1976
Jablonec nad Nisou: Museum Skla a Bižuterie, 1977, n.pag., illus.
Glass artist, engraver, sculptor.

Kalamajska-Saeed, Maryla
"Emaux sur Verre Recently Discovered in Poland"
Pantheon, 35, No. 2, April/May/June 1977, pp. 126–130, illus.
Plaques on an 18th c. tabernacle in Poland.

Kämpfer, Fritz
"Glaskunst in der DDR; Tradition und Gegenwart"
Bulletin de l'Association Internationale pour l'Histoire du Verre, 7, 1973–1976, pp. 19–24.

Klesse, Brigitte
"Eine Gruppe von Nürnberger Goldrubingläsern"
Festschrift für Waldemar Haberey
Ed. Thea Elisabeth Haevernick and Axel von Saldern
Mainz: Philipp von Zabern, 1976, pp. 65–70, illus.
17th c. gold ruby glass.

Langhamer, Antonín
"Caspar Lehman's Goblet from 1605"
Glass Review, 32, No. 1, 1977, pp. 12–15, illus.

Langhamer, Antonín
"Vladimir Linka"
Glass Review, 32, No. 6, 1977, pp. 5–9, illus.

Lüth, Erich
600 Jahre Glaserhandwerk in Hamburg
Hamburg: Hans Christians Druckerei und Verlag, 1975, 264 pp., illus.

Marek, Tejas
"Going, Going, Gone . . . Bohemian Glassware Highly Desirable"
The Collector, No. 318, Sept. 1977, p. 7+.

Prochaska, Franz
"Aus der Rechtsgeschichte des Salzburger Glaserhandwerks"
Salzburger Museum Carolino Augusteum, Jahresschrift 20, (1974), 1975, pp. 57–78.

Schieb, Gustl
"Zur Geschichte des österreichischen Glaserhandwerks"
Der Glaserlehrling, 28, No. 2, 1975, pp. 1–2.
Austrian glassworks 13th–19th c.

Schwarzwälder Glas und Glashütten
(Text by I. B. Lenzkirch)
Villingen-Schwenningen: Franziskaner-Museum
Villingen, n.d. [1976], 182 pp., illus.
Catalogue of the Franziskaner-Museum collection, with a history of Black Forest glasshouses.

Shuart, Harry Wilson
"Glass Christmas Tree Ornaments Are Getting Rarer"
Spinning Wheel, 34, No. 1, Jan.-Feb. 1978, pp. 36–38, illus.

Skrobucha, Heinz
"Rumänische Hinterglasikonen"
Die Weltkunst, 47, No. 5, March 1, 1977, pp. 380–382, illus.
Rumanian reverse-painted icons.

Štiess, Bedřich
"K dějinám zušlechťovacích závodů ve Střebli u Tachova"
Sklář a Keramik, 27, No. 2, 1977, pp. 53–54, English summary.
Observations on the history of glass refining shops at Střeble near Tachov.

Strasser, Rudolf von
Die Einschreibebüchlein des Wiener Glas- und Porzellanmalers Anton Kothgasser (1769–1851)
Karlsruhe: Corona-Verlag Karl H. Heine KG, [1977], 264 pp., illus.
Kothgasser's notebooks.

Wendt, Ralf
"Glashüttenbetrieb und Gutsherrschaft in Mecklenburg vom 17. bis 19. Jahrhundert"
Wissenschaftliche Zeitschrift der Universität Rostock, 21, 1972, Gesellschafts- und Sprachwissenschaftliche Reihe, 1, No. 1, pp. 53–80, English summary.
Development of glass factories in Mecklenburg.

ITALIAN, including *façon-de-Venise*

Bichierografia Diliniata da Giovanni Maggi
Florence: Studio Per Edizioni Scelte, 1977, Vols. I-IV, 406 pp. each vol., illus.
Drawings of glass vessels by early 17th c. artist.

CHARLESTON, R. J.
"Types of Glass Imported into the Near East and Some Fresh Examples: 15th–16th Century"
Festschrift für Peter Wilhelm Meister
Ed. Annaliese Ohm and Horst Reber
Hamburg: Dr. Ernst Hauswedell & Co., 1975, pp. 245–251, illus.

COLOGNE. KUNSTGEWERBEMUSEUM DER STADT
Kleinodien aus dem Kunstgewerbemuseum der Stadt Köln
Stuttgart: Ernst Klett, 1976, n.pag., illus.
15th c. enameled Venetian pokal.

CONTE, GIOVANNI
Tipologia e varieta' dei vetri della manna di S. Nicola
Galatina: Congedo Editore, 1976, 239 pp., illus.
Bottles, mostly 19th c.

DAVIS, FRANK
"Chinese Imitations from Meissen" (Talking About Salerooms Series)
Country Life, 161, No. 4169, May 26, 1977, pp. 1358–1359, illus.
Venetian and *façon-de-Venise* glass.

DAVIS, FRANK
"Drawings from Amsterdam" (Talking About Salerooms Series)
Country Life, 161, No. 4170, June 2, 1977, pp. 1510–1511, illus.
Façon-de-Venise 16th c. silver-mounted *latticinio* bell.

HAN, VERENA
"Retki primerak Muranskog stakla iz XVI veka otkriven na Beogradskoj tvrćavi" (Une pièce rare de verre de Murano du XVIe siècle, découverte dans la forteresse de Belgrade)
Zbornik Narodnog Muzeja, 8, 1975, pp. 481–487, illus., French summary.

HOLLISTER, PAUL
"Four Italian Nineteenth Century Millefiori Bowls at the Metropolitan"
The Glass Club Bulletin, 120, July 1977, pp. 3–11, illus.

MAEDEBACH, HEINO
"Faszination in Form und Farbe"
Kunst und Antiquitäten, 1, March/April/May 1977, pp. 12–17, illus.
Venetian and *façon-de-Venise* from the Veste Coburg collection.

OKADA, JÔ
"A Venetian Glass Ewer in the Form of a 'Nef' (A Model Ship used as a Table Centerpiece)"
Glass (Garasu), Association for Glass Art Studies, No. 1, Dec. 1975, pp. 26–29, illus., English summary.

SAN FRANCISCO. THE FINE ARTS MUSEUMS OF SAN FRANCISCO. CALIFORNIA PALACE OF THE LEGION OF HONOR
The Triumph of Humanism: A Visual Survey of the Decorative Arts of the Renaissance
San Francisco: The Fine Arts Museums of San Francisco, 1977, 95 pp., illus.
Venetian glass.

SPALLANZANI, MARCO
"Vetri, ceramiche, e oggetti metallici nella Collezione di Cosimo di Bernardo Rucellai"
Metropolitan Museum Journal, 11, 1976, pp. 137–142, illus.
Inventory of the household goods of Cosimo di Bernardo Rucellai (1470–1495).

STEPHENS, JANET
"Try This on for Size"
Rainbow Review, 7, No. 2, Feb. 1977, pp. 20–21, illus.
Venetian glass shoes.

VENICE. MUSEO VETRARIO DI MURANO
"*Vetri di Murano del '900* (text by Rosa Barovier)
Venice: Alfieri, Edizioni d'Arte, 1977, 20 pp., illus.
Brief history of Murano glass from 1895 to present

NETHERLANDISH

ANONYMOUS
"Aanwinsten Kunstnijverheid" (Craft Acquisitions)
Bulletin Museum Boymans-van Beuningen Rotterdam, 30, p. 241, illus.
17th c. bottle with calligraphic engraving by Willem van Heemskerk.

"Leerdam Unica"
Bulletin Museum Boymans-van Beuningen Rotterdam, 30, June/Aug. 1977, pp. 238–239, illus.
Review of exhibit of 50 years of Dutch glass.

"Le Musée du Verre de Liège et son action internationale depuis vingt ans"
La Vie Liégeoise, No. 12, Dec. 1977, pp. 3–17, illus.

BOONENBURG, K.
"Een huwelijksbokaal uit 1746"
Antiek, 11, No. 9, April 1977, pp. 741–747, illus.
Engraved pokal.

DAVIS, FRANK
"An Uncompromising Pre-Raphaelite" (Talking About Salerooms Series)
Country Life, 162, No. 4194, Nov. 17, 1977, pp. 1466–1467, illus.
Dutch bottle with calligraphic engraving, about 1680.

DÜSSELDORF. KUNSTMUSEUM
Leerdam unica; 50 Jahre modernes niederländisches Glas. 50 jaar modern nederlands glas. [ausstellung]
Düsseldorf: Kunstmuseum, 1977, 148 pp., illus.

LIÈGE. MUSÉE DE L'ART WALLON.
"*Le siècle de Louis XIV au pays de Liège (1580–1723)*
Liège: Musée de l'Art Wallon, 1975, 214 pp., illus.
Glass, pp. 181–184, by Joseph Philippe.

VAN HELDEN, ALBERT
The Invention of the Telescope
Philadelphia: The American Philosophical Society Transactions, 1977, 67 pp., illus.
Possible lens origin at Dutch glassworks, Middleburg.

NORTH AMERICAN
United States – Canada – Mexico

ANONYMOUS
"Corning Glass Collection on View in Major Exhibition"
Antique Monthly, 10, No. 7, May 1977, p. 16A, illus.

"18th Century Glassmakers: Four Efficient Germans"
The Heritage Newsletter, 8, No. 10, Dec. 1977, p. 6.
Wistar, Stiegel, Amelung, Stanger.

"Glass Companies in Toledo"
Bulletin, Glass Collectors Club of Toledo, Nov. 1977, p. 3.

"Glass Reproductions Can Be Spotted with a Little Effort and Study"
The Collector, 8, No. 317, Aug. 1977, p. 6.

"Hatpin Holders"
The Antiques Journal, 32, No. 12, Dec. 1977, p. 37, illus.
Brief article about cut glass holders.

"An Important Accession with a Unique History"
The Acorn, 11, No. 2, May 1977, p. 6, illus.
Sandwich dolphin stem candlesticks.

"Jamestown, Va. Glasshouse (1608) Reconstructed"
American Antiques, 4, No. 9, Sept. 1976, p. 39, illus.

"Made in Corning: Glass"
Americana, 5, No. 2, May/June 1977, pp. 56–57, illus.

"Pattern of the Month"
Continuing series in *National Depression Glass Journal*, 5, No. 10, Jan. 1977–6, No. 9, Dec. 1977.

"Steuben; Seventy Years of American Glassmaking"
American Antiques, 4, No. 9, Sept. 1976, p. 34, illus.

"Unique Insights into Life at the Glass Factory; from Barnstable Journal 1839"
The Acorn, 11, No. 3, Aug. 1977, p. 5.
Sandwich Co.

ALLEN, OLIVE
"Antiques; Pressed Glass"
Collectors News, 18, No. 6, Oct. 1977, p. 52, illus.

ALLISON, GRACE
"A Glass Closeup"
Rainbow Review, 7, No. 9, Sept. 1977, p. 12, illus.

ALLISON, GRACE
Kemple Glass Works 1945–1970
Costa Mesa, Calif.: Rainbow Review Glass Journal, 1977, 14 pp., illus.

ANGUS, FERN
"Fern Looks at Depression Glass"
Irregular series in *The Collector*, V. 8, No. 310, Jan. 1977–V. 9, No. 321, Dec. 1977.

ANGUS, FERN
"Now and Then"
Irregular series in *National Depression Glass Journal*, 5, No. 10, Jan. 1977–6, No. 9, Dec. 1977.

The Antique Trader Weekly Annual of Articles on Antiques; Volume VI
Reprints of articles which appeared in *The Antique Trader Weekly* from July 1976 to March 1977.

ARCHER, MARGARET AND DOUGLAS
Glass Candlesticks; Book 2
Paducah, Ky.: Collector Books, 1977, 110 pp., illus.

The Art of Making Fine Glassware
Cambridge, Ohio: The Cambridge Glass Co., 1939, 32 pp., illus.

BAKER, HENRY
"Archaeological Investigations at Fort Cooper, Inverness, Florida"
Bureau of Historic Sites and Properties, Bulletin No. 5, 1976, Florida Department of State, pp. 21–45, illus.
Bottle fragments, window glass, etc., from a 19th c. trash pit.

Bakewell, Pears & Co. Glass Catalogue [c. 1875] (reprint)
Pittsburgh, Pa.: Thomas C. Pears III, 1977, 49 pp., illus.

BARNETT, JERRY
"The Paden City Party Line"
Irregular series in *National Depression Glass Journal*, 5, No. 10, Jan. 1977–6, No. 9, Dec. 1977.

BELL, BETTY
"Bits and Pieces"
Continuing series in *National Depression Glass Journal*, 5, No. 10, Jan. 1977–6, No. 9, Dec. 1977.

BELL, BETTY
"Glass Figurals"
Continuing series in *Depression Glass Daze*, 6, No. 11, Jan. 1, 1977–7, No. 10, Dec. 1977.

BLOESSER, JIM
"Where Is It?"
Continuing series in *Rainbow Review*, 6, No. 8, Aug. 1976–7, No. 1, Jan. 1977.
Six articles on the major depression glass companies.

BLOESSER, ROBERT
"Glassware"
Rainbow Review, 7, No. 4, April 1977, pp. 28–29, illus.

"Manhatten Revisited"
Rainbow Review, 7, No. 3, March 1977, p. 38, illus.

"The Pre-Hocking 'Manhatten'"
Rainbow Review, 7, No. 2, Feb. 1977, pp. 33–35, illus.

BOGGESS, BILL AND LOUISE
American Brilliant Cut Glass
New York: Crown Publishers, 1977, 183 pp., illus.

BOGGESS, LOUISE
"Fabulous Libbey Cut Glass"
The Collector, 8, No. 314, May 1977, pp. 1+, illus.
Part II: *The Collector*, 8, No. 315, June 1977, p. 23, illus.

BOGGESS, LOUISE
"Glass Reflections"
Irregular series in *The Collector*, V. 8, No. 310, Jan. 1977–V. 9, No. 321, Dec. 1977.

BOLKEN, KANDYCE SMITH
"Carnival Glass"
National Depression Glass Journal, 6, No. 3, June 1977, p. 24, illus.

"Glass Hats"
National Depression Glass Journal, 5, No. 12, March 1977, p. 30, illus.

"Glass Shoes"
Depression Glass Daze, 7, No. 1, March 1977, p. 32, illus.

"Kollecting with Kandi"
Irregular series in *Rainbow Review*, 7, No. 1, Jan. 1977–7, No. 12, Dec. 1977.

"Slag Glass"
National Depression Glass Journal, 6, No. 2, May 1977, p. 24.

"Sun Colored Glass"
National Depression Glass Journal, 6, No. 1, April 1977, p. 8, illus.

"Update on Sun Colored Glass"
National Depression Glass Journal, 6, No. 4, July 1977, p. 20, illus.

BOURQUE, JOSEPH
"Floral Shakers"
Depression Glass Daze, 7, No. 2, April 1977, p. 31, illus.

"Important Discovery on Early Akro History"
Depression Glass Daze, 7, No. 8, Oct. 1977, pp. 38–39, illus.

"Variegated Akro"
Depression Glass Daze, 7, No. 3, May 1977, pp. 32–33, illus.

BOYD, RALPH AND LOUISE
Greentown
Des Moines, Iowa: Wallace-Homestead Book Co., 1972, n.pag., illus.

BRADLEY, STEPHEN H.; RYAN, C. S.; AND RYAN, R. R.
Heisey Stemware
Newark, Ohio: Spencer Walker Press, Inc., 1976, 238 pp., illus.

Bredehoft, Neila
"Heisey Decorations"
Continuing series in *Heisey News*, 6, No. 1, Jan. 25, 1977–6, No. 12, Dec. 25, 1977.

Bredehoft, Tom
"Tumblers"
Continuing series in *Heisey News*, 6, No. 1, Jan. 25, 1977–6, No. 12, Dec. 25, 1977.

Brockton, Mass. Brockton Art Center–Fuller Memorial
Artful Toil: Artistic Innovation in an Age of Enterprise
Brockton: Brockton Art Center, 1977, 31 pp., illus.
Includes 19th c. glass items.

Butterfield, Oliver
"Marble Collecting for Keeps"
Yankee, Sept. 1977, pp. 174–184, illus.

Cheesman, George
"Frederick County's Forgotten Glassmaker"
Maryland, 9, No. 4, Summer 1977, pp. 27–31, illus.

Collins, Alma
"Degenhart Update"
Irregular series in *Rainbow Review*, 7, No. 1, Jan. 1977–7, No. 12, Dec. 1977.
John Frederick Amelung.

Consentino, Geraldine and Stewart, Regina
Carnival Glass; A Guide for the Beginning Collector
New York: Golden Press, 1976, 128 pp., illus.

Cooke, Lawrence, ed.
Lighting in America; from Colonial Rushlights to Victorian Chandeliers
New York: Main Street/Universe Books, 1977, 159 pp., illus.
Reprints from articles in *Antiques* 1924–1972.

Cotter, John
"Current Research, Northeast"
The Society for Historical Archaeology Newsletter, 10, No. 2, June 1977, pp. 12–21.
Glass finds at Middletown, Conn. and Deerfield, Mass.

Courter, J. W.
"How to Tell New Lincoln Drape Lamps from Original Aladdins"
National Depression Glass Journal, 6, No. 4, July 1977, pp. 38–39, illus.

Courter, J. W.
"Light on . . . Aladdin"
Irregular series in *Rainbow Review*, 7, No. 1, Jan. 1977–7, No. 12, Dec. 1977.

Davidson, Marshall
New York: A Pictorial History
New York: Charles Scribner's Sons, 1977, 357 pp., illus.
Crystal Palace of New York City; Corning Glass Works.

The Daze Past—1976
Otisville, Mich.: Nora Koch, 1977, a compilation of articles from *Depression Glass Daze* during 1976, Nora Koch, comp., n.pag., illus.

Dellandrea, Jon
"Bottles in Canada"
Canadian Collector, 12, No. 5, Sept./Oct. 1977, pp. 26–29, illus.

DeLorme, Rita
"Advertising Ashtrays of the Thirties"
Depression Glass Daze, 7, No. 5, July 1977, p. 36, illus.

DeLorme, Rita
"Just a Few Mug Shots"
Depression Glass Daze, 7, No. 2, April 1977, p. 29, illus.
Miniature mugs.

Dickerson, Beverly
"Glass Tumblers"
American Collector, 8, No. 3, March 1977, pp. 8–9, illus.

Dimitroff, Thomas and Janes, Lois
History of the Corning Painted Post Area: 200 Years in Painted Post Country
Corning, N.Y.: Corning Area Bicentennial Committee, 1977, 384 pp., illus.
Corning cut glass firms, Frederick Carder, Steuben, etc.

Dipboye, Marilyn
"End-of-Day Glass"
American Collector, 8, No. 6, June 1977, pp. 23+, illus.

Dipboye, Marilyn
"Lamps by Handel"
American Collector, 8, No. 5, May 1977, pp. 23+, illus.

Doyle, Jim
"Did You Know?"
Irregular series in *National Depression Glass Journal*, 5, No. 10, Jan. 1977–6, No. 9, Dec. 1977.

Drager, Heloise
"The Lure of Ruby-Stained Glass"
American Collector, 8, No. 7, July 1977, pp. 18–19, illus.

Eastwood, Maud
The Antique Doorknob
Forest Grove, Oregon: Maudie Eastwood, 1976, 206 pp., illus.

Evanoff, Betty
"Lightning Rod Balls"
American Collector, 8, No. 1, Jan. 1977, pp. 12–13+, illus.

Farrar, Estelle Sinclaire
"Cut and Engraved Glass of Corning, N.Y., 1868–1940"
American Antiques, 5, No. 5, May 1977, pp. 25–28, illus.

Farrar, Estelle Sinclaire
"The Master Engravers of Corning, New York"
Antiques, 112, No. 4, Oct. 1977, pp. 726–731, illus.

Fauster, Carl
"Brilliant Period Cut Glass; Part II"
Antiques Gazette, 2, No. 12, July 1976, pp. 2–3, illus.

Fauster, Carl
"The McKinley Punch Bowl"
American Antiques, 4, No. 9, Sept. 1976, pp. 35–38, illus.

Fichera, Lou and Perkins, Ron
"Pairpoint and Handel Lamps"
Program brochure for National Antiques Show, Madison Square Garden, New York, 1977, p. 8, illus.

Florence, Jr., Gene
"Depression Glass"
Rainbow Review, 7, No. 8, Aug. 1977, p. 24, illus.
U.S. Glass Company.

Frazier, Dale
"Lightning Rod Balls"
Old Bottle Magazine, 10, No. 3, March 1977, pp. 14–17, illus.

Gammon, Don
Heisey Orchid Etch Glassware
Cape Girardeau, Mo.: Don Gammon, n.d., 20 pp., illus.

Gardner, Paul
American Glass
Washington, D.C.: Smithsonian Institution Press, 1977, 12 pp., illus.

Garrett, Jerry D.
"Digging in the Past"
Continuing series in *National Greentown Glass Association Newsletter*, 4, No. 1, Jan. 1977–4, No. 4, Oct. 1977.
Indiana Tumbler and Goblet Co. history.

Gibbs, W. H. & Co.
Cut Glass (catalog)
Stroudsburg, Pa.: W. H. Gibbs & Co., n.d., [1914–1935], n.pag., illus.

Gibbs, W. H. & Co.
Rich Cut Glass (catalog)
Stroudsburg, Pa.: W. H. Gibbs and Co., n.d., [1914–1935], n.pag., illus.

Gibbs, W. H. & Co.
Rich Cut Glass (catalog)
Stroudsburg, Pa.: W. H. Gibbs and Co., n.d., [1914–1935], n.pag., illus.

Gibbs, W. H. & Co.
Scatter Sunshine at Half Price (catalog)
Stroudsburg, Pa.: W. H. Gibbs & Co., n.d., [1914–1935], n.pag., illus.
Cut glass tableware.

Gores, Stan
"Centennial Paperweights 100 Years Later"
Glass Paperweight Symposium, June 1976, The John Nelson Bergstrom Art Center and Museum, Neenah, Wisc., pp. 17–24.

Gustafson, Eleanor, ed.
"Clues and Footnotes"
Antiques, 112, No. 4, Oct. 1977, p. 753.
Spun-glass fabric for dresses, 1897.

Gustafson, Eleanor
"Museum Accessions"
Antiques, 112, No. 6, Dec. 1977, pp. 1084–1088, illus.
Wistarburg blown glass taper stick at Winterthur.

Hall, Donald
"The Why and When of Depression Glass"
The New York–Pennsylvania Collector, 2, No. 7, Sept. 1977, pp. 4–5, illus.

Handler, Sandra
"Collecting Depression Glass in Canada"
Rainbow Review, 7, No. 4, April 1977, pp. 34–35, illus.

Handler, Sandra
"Insights into Tableware; The Dominion Glass Company of Canada"
Depression Glass Daze, 6, No. 12, Feb. 1977, p. 41, illus.

Hartley, Julia and Cobb, Mary
Early American Pattern Glass
[Lubbock, Texas?]: Julia Hartley and Mary Cobb, 1976, 88 pp., illus.

Heacock, William
Encyclopedia of Victorian Colored Pattern Glass, Book 4: Custard Glass from A to Z
Marietta, Ohio: Antique Publications, 1976, 68 pp., illus.

Heacock, William
"Glimpses at Old Glass"
Irregular series in *Rainbow Review*, 7, No. 1, Jan. 1977–7, No. 12, Dec. 1977.

Heissenbuttel, Orva
"Speaking of Glass . . ."
Irregular series in *Rainbow Review*, 7, No. 1, Jan. 1977–7, No. 12, Dec. 1977.

Historical Society of Temple, New Hampshire
A History of Temple, New Hampshire 1768–1976 (Anne Lunt and others, eds.)
Temple: Historical Society of Temple, N. H., 1976, 906 pp., illus.
Temple Glass Works pp. 211–219.

Hollister, Paul
"Antique American Paperweights"
Acquire, 5, No. 1, March 1977, pp. 28–29, illus.

Horton, Christine
"Giants in Glass Houses; American Cut Glass 1880–1915: The Brilliant Period"
American Antiques, 4, No. 9, Sept. 1976, pp. 26–33, illus.

Horton, Christine
"Patterns, Patents and Trademarks of Cut and Engraved Glass; Part I"
American Antiques, 5, No. 5, May 1977, pp. 18–24, illus.

Horton, Christine
"Patterns, Patents and Trademarks of Cut and Engraved Glass; Part II"
American Antiques, 5, No. 8, Aug. 1977, pp. 21–27, illus.

Howard, Marjorie
"Fostoria's Navarre Pattern"
National Depression Glass Journal, 6, No. 6, Sept. 1977, pp. 22–23, illus.

Huffman, George
"New Finds in Fragments: Beehive"
National Greentown Glass Association Newsletter, 4, No. 1, Jan. 1977, pp. 3–4, illus.

Huffman, George
"New Finds in Fragments: Flatiron"
National Greentown Glass Association Newsletter, 4, No. 2, April 1977, pp. 5–6, illus.

Iroquois Door Company; Catalog No. 415
Buffalo, N.Y.: Iroquois Door Co., n.d. [1915?], 207 pp., illus.
Door lights: leaded clear and art glass designs in original catalog.

Johnson, Mary Ellen and Chapman, Bernadine
"Stretch Glass"
American Collector, 8, No. 4, April 1977, pp. 12–13, illus.

Johnson, Susan C.
"150 Years of American Glass"
Carnegie Magazine, 51, No. 7, Sept. 1977, pp. 13–18, illus.

Judd, Cameron
"Butter Molds: True Folk Art"
The Antiques Journal, 32, No. 3, March 1977, pp. 18–21, illus.
19th c. glass molds.

Kinch, Blanche
"A History of American Glass on Exhibit at The Sandwich Glass Museum"
American Antiques, 4, No. 9, Sept. 1976, pp. 23–25, illus.

Kinne, Ann Spencer
"Connecticut Pattern Glass"
Spinning Wheel, 33, No. 1, Jan./Feb. 1977, pp. 22–24, illus.

Kinne, Ann Spencer
"Hobbs, Brockunier & Company's Opalescent Pattern #336"
Spinning Wheel, 33, No. 10, Dec. 1977, p. 22, illus.

Kinne, Ann Spencer
"The Nickel Plate Glass Company's Opalescent Wares"
Spinning Wheel, 33, No. 8, Oct. 1977, pp. 26–27, illus.

KRAUSE, GAIL
"Delightfully Duncan"
Irregular series in *Rainbow Review*, 7, No. 1, Jan. 1977–7, No. 12, Dec. 1977.

KRAUSE, GAIL
A Pictorial History of Duncan & Miller Glass
Washington, Pa.: Gail Krause, 1976, 32 pp., illus.

LASHER, FAITH
"The Storybook Mug Story"
American Collector, 8, No. 11, Nov. 1977, pp. 20–21, illus.

LECHLER, DORIS
Children's Glass Dishes
Nashville, Tenn.: Thomas Nelson, Inc., 1976, 224 pp., illus.

LEHNER, LOIS
"From the Desk of Lois Lehner"
Continuing series in *Depression Glass Daze*, 6, No. 11, Jan. 1, 1977–7, No. 10, Dec. 1977.

LEHNER, LOIS
"Glass Notes"
Irregular series in *Rainbow Review*, 7, No. 1, Jan. 1977–7, No. 12, Dec. 1977.

LEHNER, LOIS
"Slag Glass"
Depression Glass Daze, 7, No. 4, June 1977, p. 1+, illus.

LIPPERT, CATHERINE AND MEASELL, JAMES
"Greentown Glass: Indiana Tumbler and Goblet Company"
Antiques, 111, No. 4, April 1977, pp. 774–781, illus.

LOKAY, JOSEPH
"Cut Glass: The J. Hoare & Company"
Rainbow Review, 7, No. 3, March 1977, pp. 16–17, illus.

LOKAY, JOSEPH
"Heisey Patterns and Patents"
Continuing series in *Heisey News*, 6, No. 1, Jan. 25, 1977–6, No. 12, Dec. 25, 1977.

LOKAY, JOSEPH
"Westmoreland Glass"
Rainbow Review, 7, No. 5, May 1977, pp. 30–31, illus.

LOVELL, RUSSELL, JR.
"Sandwich in Its Heyday; Part II"
The Acorn, 11, No. 2, May 1977, pp. 3–4.

MCCORMICK, FRANCES
"Reamers"
Depression Glass Daze, 7, No. 10, Dec. 1, 1977, p. 32, illus.

MARSH, DICK
"Heisey Animals—Figurines"
Heisey News, 6, No. 2, Feb. 25, 1977, p. 2.

MEASELL, JAMES
"Diamond Prisms"
National Greentown Glass Association Newsletter, 4, No. 4, Oct. 1977, p. 4, illus.

METCALF, JACK
"Heisey Candlesticks"
Continuing series in *Heisey News*, 6, No. 1, Jan. 25, 1977–6, No. 12, Dec. 25, 1977.

MILLER, PAUL B. ed.
"A History of the Tiffin Glass Company, Tiffin, Ohio"
Stretch Glass Society Newsletter, 3, No. 4, April 1977, pp. 5–8.

MILLER, PAUL B. ed.
"History of the Tiffin Glass Company, Tiffin, Ohio, Con't."
Stretch Glass Society Newsletter, 4, No. 2, Oct. 1977, pp. 4–6.

MILLER, PAUL B. ed.
"Sweeney Punch Bowl"
Stretch Glass Society Newsletter, 3, No. 3, Jan. 1977, p. 9.
1884, Wheeling, W. Va.

MIZERACK, BARBARA
"Stacks of Facts"
Irregular series in *Depression Glass Daze*, 6, No. 11, Jan. 1977–7, No. 10, Dec. 1977.
U.S. glass factories.

MONAGHAN, NANCY
"Stretch Glass: Carnival's Kissing Cousin"
Stretch Glass Society Newsletter, 4, No. 1, July 1977, pp. 6–7.

MORSE, SYLVIA
"Stiegel Glass: A Gift of Love"
Old Bottle Magazine, 10, No. 9, Sept. 1977, pp. 12–13, illus.

MULLIKIN, SEAN
"Glass Making by the (Borrowed) Book"
Old Bottle Magazine, 10, No. 8, Aug. 1977, pp. 10–13+, illus.
History of the Illinois Glass Co., precursor of Owens-Illinois.

MUNCIE, INDIANA. BALL STATE UNIVERSITY ART GALLERY
Indiana Pressed Glass from the Gas-Boom Era
Muncie, Ind.: Ball State University, 1977, n.pag., illus.

MYERS, SARA
"Experimental Patterns Come to Light"
Depression Glass Daze, 6, No. 11, Jan. 1977, p. 1+, illus.
Hazel-Atlas glass.

NEENAH, WISCONSIN. JOHN NELSON BERGSTROM ART CENTER AND MUSEUM
Thirty-five Years of New England Area Glass: Cut, Mold-Blown, and Lacy 1815–1850
Neenah: John Nelson Bergstrom Art Center and Museum, 1977, n.pag.

NOËL HUME, IVOR
Archaeology and Wetherburn's Tavern
Williamsburg, Va.: The Colonial Williamsburg Foundation, 1969, 47 pp., illus.
Early 18th c. bottles, wineglass stems, decanter stoppers, etc.

NOËL HUME, IVOR
Williamsburg Cabinetmakers; The Archaeological Evidence
Williamsburg, Va.: The Colonial Williamsburg Foundation, 1971, 48 pp., illus.
Fragments of mirror plate, some engraved.

OLIVER, ELIZABETH
"Look to Your Attic! Portland Glass is Coming Back"
Down East Magazine, April 1977, pp. 52–53, illus.
Portland Glass Co. 1864–1873.

PAPAPANU, SOPHIA
"Fostoria's Heather, Camelia and Kamak"
Depression Glass Daze, 7, No. 6, Aug. 1, 1977, p. 38, illus.
"Houze Lamps"
Depression Glass Daze, 7, No. 5, July 1977, p. 32, illus.

"Houze Rocker Blotters"
Depression Glass Daze, 7, No. 10, Dec. 1, 1977, p. 1+, illus.
Glass figural blotters.

"Powder Jars"
Depression Glass Daze, 6, No. 12, Feb. 1977, p. 35, illus.

PEARSON, J. MICHAEL
Encyclopedia of American Cut and Engraved Glass 1880–1917; Volume II, Realistic Patterns
Miami Beach Fla.: J. Michael Pearson, 1977, 172 pp., illus.

PETERSON, MARTY
"Who Wants Old Light Bulbs?"
American Collector, 8, No. 2, Feb. 1977, pp. 12–13, illus.

PFEIFFER, EDDIE
"Little Known Facts about the Window Glass Factory"
Heritage Newsletter, 8, No. 8, Oct. 1977, p. 7.
New Jersey factory.

PFEIFFER, MARY LOU
"Milk Glass"
Antique Bottle World, 3, No. 10, Feb. 1977, p. 31.

REAM, LOUISE
"Hawthorne"
Heisey News, 6, No. 5, May 25, 1977, p. 12, illus.

REAM, LOUISE
"More on Verlys by Heisey"
Heisey News, 6, No. 7, July 25, 1977, pp. 16–18, illus.

REAM, LOUISE; BREDEHOFT, NEILA AND BREDEHOFT, THOMAS
Encyclopedia of Heisey Glassware, Volume 1
Newark, Ohio: Heisey Collectors of America, Inc., 1977, 228 pp., illus.

REDINGER, CAROLYN
"The Elusive Powder Bowl"
Irregular series in *Depression Glass Daze*, 6, No. 11, Jan. 1, 1977–7, No. 10, Dec. 1977.

ROADS, JACK
"Jack Roads on Glass"
Irregular series in *Rainbow Review*, 7, No. 1, Jan. 1977–7, No. 12, Dec. 1977.

ROSSEN, HOWARD AND KADUCK, JOHN
Columbian World's Fair Collectibles; Chicago (1892–1893)
Des Moines, Iowa: Wallace-Homestead Book Co., 1976, 149 pp., illus.

THE RUSHLIGHT CLUB, comp.
Patented Lighting
n.l.: The Rushlight Club, 1976, 101 pp., illus.
American lamps and lanterns.

ST. PETERSBURG, FLORIDA. MUSEUM OF FINE ARTS
"The New Vision": American Styles of 1876–1910
St. Petersburg: Museum of Fine Arts, 1976, n.pag., illus.
Glass of the period on loan from The Metropolitan Museum of Art.

SCHALLER-HOERR CO.
Guaranteed Millwork and Building Materials: Catalogue No. 98
Chicago: Schaller-Hoerr Co., 1909, 208 pp., illus.
Leaded art glass, mitered and beveled plate glass, leaded "crystal" glass.

SCHROEDER, BILL
Carnival Glass; Collector's Illustrated Price Guide
Paducah, Ky.: Collector Books, 1977, 63 pp., illus.

SCHROEDER, BILL
Cut Glass; Collector's Illustrated Price Guide
Paducah, Ky.: Collector Books, 1977, 63 pp., illus.

SCOTT, VIRGINIA
"Glass Ad Search and Research"
Irregular series in *Depression Glass Daze*, 6, No. 11, Jan. 1977–7, No. 10, Dec. 1977.

SCOTT, VIRGINIA
Imperial Candelwick
Athens, Georgia: Virginia R. Scott, 1975, Imperial Glass Corporation, Bellaire, Ohio
Reprint catalogs, advertisement, brochures of 1930's–1940's, n.pag., illus.

SCOTT, VIRGINIA
"Looking through the Ads"
Continuing series in *Rainbow Review*, 7, No. 1, Jan. 1977–7, No. 12, Dec. 1977.

SEARS, ROEBUCK AND CO.
Lighting Fixtures by Sears (catalog)
n.l.: Sears, Roebuck and Co., 1934, 32 pp., illus.

SELL, VIRGINIA
"Knuckle Down"
Rainbow Review, 7, No. 9, Sept. 1977, p. 13.
Marbles.

SELL, VIRGINIA
"Master Glass Company"
Depression Glass Daze, 7, No. 4, June 1, 1977, p. 33.
Clarksburg, W. Va. factory.

SERRA, ANNE
"Plain Charm, or the Story of Blown-Molded Glass"
The New York–Pennsylvania Collector, 2, No. 9, Nov. 1977, pp. 10–11+, illus.

SHEPHERD, ELIZABETH
"Canadian Glass Collecting"
Canadian Collector, 12, No. 4, July/Aug. 1977, pp. 45–47, illus.

SHERRILL, SARAH
"Current and Coming: American Glass"
Antiques, 111, No. 6, June 1977, pp. 1088+, illus.
Cut and engraved glass show at Corning.

SHUMPERT, GWEN
"Gwen's Glassline"
Continuing series in *Rainbow Review*, 7, No. 1, Jan. 1977–7, No. 12, Dec. 1977.

SIONAKIDES, GEORGE
"Duncan's 'Spiral Flutes'"
Depression Glass Daze, 7, No. 8, Oct. 1, 1977, p. 30, illus.

SIONAKIDES, GEORGE
"Frances Pattern Update"
Depression Glass Daze, 7, No. 9, Nov. 1, 1977, p. 41, illus.

SLAYTON, IRIS
"The Cherry Blossom Cookie Jar"
National Depression Glass Journal, 6, No. 5, Aug. 1977, p. 15, illus.

SMALL, DANIEL
"Her Name Is Mrs. Button"
American Collector, 8, No. 11, Nov. 1977, pp. 18–19, illus.
Glass buttons.

SMITH, WILLIAM
"Collecting Kerosene Lanterns"
Spinning Wheel, 33, No. 4, May 1977, pp. 24–25, illus.

Solar Prisms in Ornamental Designs (catalog)
Cleveland, Ohio: The Solar Prism Co., n.d., [c. 1920], n.pag., illus.

Sparacio, Carl
"Stems"
Continuing series in *Heisey News*, 6, No. 1, Jan. 25, 1977–6, No. 12, Dec. 25, 1977.

Sperling, Patricia
"Captivating Carnival Glass"
Wonderful World of Ohio, Feb. 1968, pp. 8–13, illus.

Spillman, Jane and Farrar, Estelle Sinclaire
The Cut and Engraved Glass of Corning, 1868–1940
Corning, N.Y.: The Corning Museum of Glass, 1977, 101 pp., illus.

Springer, Lynn
"19th Century New England Glass"
Saint Louis Art Museum Bulletin, 11, No. 1, Jan.–Feb. 1975, pp. 11–13, illus.
Several acquisitions, including a Louis Vaupel goblet of 1875–80.

Stanton, Joyce
""Duncan's Lily of the Valley"
Depression Glass Daze, 7, No. 8, Oct. 1977, p. 35, illus.

Stearns, Glenita
"Glancing Back"
Continuing series in *Rainbow Review*, 7, No. 1, Jan. 1977–7, No. 12, Dec. 1977.
"Imperial's American Beauty Rose"
Depression Glass Daze, 7, No. 3, May 1977, p. 28, illus.
"Minuet by Heisey"
Depression Glass Daze, 7, No. 9, Nov. 1, 1977, p. 38, illus.
Princess Feather by Westmoreland"
Depression Glass Daze, 7, No. 3, May 1977, p. 1, illus.
"Tally-Ho"
Depression Glass Daze, 7 No. 8, Oct. 1977, p. 31, illus.
"Westmoreland's Princess Feather"
Depression Glass Daze, 7, No. 6, Aug. 1, 1977, p. 30, illus.

Stephens, Janet
"Try This on for Size"
Continuing series in *Rainbow Review*, 7, No. 1, Jan. 1977–7, No. 12, Dec. 1977.
Glass shoes and boots.

Stout, Sandra McPhee
Depression Glass III
Des Moines, Iowa: Wallace-Homestead Book Co., 1976, 44 pp., illus.

Thompson, Frances
"Pattern Glass"
Irregular series in *Rainbow Review*, 7, No. 1, Jan. 1977–7, No. 12, Dec. 1977.

Troutman, Robert S.
"The Madison Pattern in Pressed Glass"
Spinning Wheel, 33, No. 9, Nov. 1977, p. 14, illus.

Van Den Hogen, Eileen
"New England Paperweights"
The Antiques Journal, 32, No. 4, April 1977, pp. 12–14, illus.

Van Pelt, Mary
"Fantastic Figurines"
Irregular series in *National Depression Glass Journal*, 5, No. 10, Jan. 1977–6, No. 9, Dec. 1977.

Van Pelt, Mary
"Fantastic Figurines"
Continuing series in *Rainbow Review*, 7, No. 1, Jan. 1977–7, No. 12, Dec. 1977.

Vogel, Clarence
Heisey's Art and Colored Glass 1922–1942; Book III
Plymouth, Ohio: Heisey Publications, 1970, 183 pp., illus.
Heisey's Colonial Years: A Picture Guide to Heisey Glass 1906–1922; Book II
Plymouth, Ohio: Heisey Publications, 1969, 173 pp., illus.
Heisey's Early and Late Years 1896–1958; Book IV
Plymouth, Ohio: Heisey Publications, 1971, 168 pp., illus.
Heisey's First Ten Years 1896–1905; Book I
Plymouth, Ohio: Heisey Publications, 1969, 132 pp., illus.
"Heisey's Glassware"
Continuing series in *Rainbow Review*, 7, No. 1, Jan. 1977–7, No. 12, Dec. 1977.

Wetz, John
"The Horse Railroad"
The Acorn, 11, No. 3, Aug. 1977, pp. 1–3, illus.
"Success to the Railroad" flask and a possible relationship to the Sandwich Co. horse railway.

Wheat, R. E.
"Fostoria on My Mind"
Continuing series in *National Depression Glass Journal*, 5, No. 10, Jan. 1977–6, No. 9, Dec. 1977

Wilson, Kenneth M.
"American Paperweights; Origins and Influences"
Glass Paperweight Symposium, June 1976, The John Nelson Bergstrom Art Center and Museum, Neenah, Wis., pp. 42–46.

Wollett, Mary and Bill
"American Historical Glass"
Continuing series in *Hobbies*, 81, No 11, Jan. 1977–82, No. 10, Dec. 1977.

Worcester, Mass.
Worcester Art Museum
The Early Republic: Consolidation of Revolutionary Goals (text by Stephen B. Jareckie)
Worcester: Worcester Art Museum, 1976, 99 pp., illus.
Cut and blown-mold tableware, 19th c.

Worcester, Mass.
Worcester Art Museum
The Second Fifty Years: American Art 1826–1876 (Text by Timothy Riggs)
Worcester: Worcester Art Museum, 1976, 95 pp., illus.
Dip-molded and early pressed pieces.

Bottles, Flasks, Jars, and Insulators

Anonymous
"Bininger Bottles Favored by Collectors"
Bottle News, 6, No. 59, Nov. 1977, pp. 10–11, illus.
"History Told in Bottles"
Heritage Newsletter, 8, No. 1, Jan. 1977, p. 1+, illus.
Brief item about Owens-Illinois bottle exhibit.
"Mineral Delight! Saratogas"
Bottle News, 6, No. 58, Oct. 1977, pp. 20–21, illus.
"Owens-Illinois Bottle Exhibit"
The Federation Letter, 7, No. 3, March 1977, pp. 10–13, illus.

Baldwin, Joseph and Wagner, James
"The Bottles of Dr. John O. Hill"
The Federation Letter, 7, No. 2, Feb. 1977, pp. 12–17, illus.

Bickerton, L. M.
"Glass Bottles: The Collection of a Wine-Merchant"
The Australasian Antique Collector, 17, 1976, pp. 70–73, illus.
Walter A. Sichel collection.

Blasi, Betty
A Bit about Balsams; A Chapter in the History of Nineteenth Century Medicine
Louisville, Ky.: Farley-Goepper Co., 1974, 175 pp., illus.

Boggs, W. Doris
"Jumbo: 'Good Enuf For Me'"
The Antiques Journal, 32, No. 2, Feb. 1977, pp. 37–39, illus.
Jumbo peanut butter jars.

Bologna, Sando
"Barber Bottles"
The Antiques Journal, 32, No. 11, Nov. 1977, pp. 16–18, illus.

Boster, Curt
"The Tie Wire"
Irregular series in *Antique Bottle News*, 3, No. 9, Jan. 1977–4, No. 7, Dec. 1977.
Insulators.

The Brookfield Glass Co. Insulators: Catalog No. 51; 1912
Sequim, Washington: Marion Milholland, 1974, reprint, 126 pp., illus.

Brown, Gerald
The Story of Fred M. Locke and His Insulators
Two Buttes, Col.: Gerald Brown, 1977, 82 pp., illus.

Brown, Roy
"Collectors Agree: Hail Columbia!"
Antique Bottle World, 4, No. 6, Nov. 1977, pp. 6–7, illus.
Columbia flasks.

Brown, Roy
"A Hungarian Patriot's Story: History in Glass"
Antique Bottle World, 4, No. 4, Sept. 1977, pp. 6–9, illus.
Louis Kossuth flasks.

Cannon, Richard
"Those Scarce Southern Sarsaparillas"
Old Bottle Magazine, 10, No. 11, Nov. 1977, pp. 6–7, illus.

Creswick, Alice
The Red Book of Fruit Jars No. 3
Grand Rapids, Mich.: Alice M. Creswick, 1977, 224 pp., illus

Davis, Chris
"Read This — It's Essential"
Antique Bottle World, 4, No. 7, Dec. 1977, pp. 14–16, illus.
Oil of peppermint bottles.

Delling, David and Marilyn
Before Threads
n.l.: David and Marilyn Delling, 1971, 49 pp., illus.
Insulators, c. 1845–1870.

Eichhorn, Gary
"Roll out the Barrels"
Old Bottle Magazine, 10, No. 8, Aug. 1977, pp. 6–8, illus.

Fischer, John
"B.M.G. Co. Embossings"
Insulators, 9, No. 9, Nov. 1977, pp. 7–8.

Fountain, John C.
"John C. Fountain's Scrapbook"
Continuing series in *Bottle News*, 5, No. 49, Jan. 1977–6, No. 60, Dec. 1977.

Fredgant, Don
"The Leonardi Bottles: History Hiding in Glass"
Bottle News, 6, No. 60, Dec. 1977, pp. 18–19, illus.

Goldberg, Al
"Of Flasks and Bottles"
The Blade (Toledo) Sunday Magazine, Sept. 12, 1976, pp. 4–11, illus.

Hayes, Jack
"Early Western Canadian Telegraph Lines and Insulators"
Insulators, 9, No. 9, Nov. 1977, pp. 3–6, illus.

Herberta, Victoria
"'Sodalicious' or How It All Began"
Collector News, 18, No. 6, Oct. 1977, p. 39, illus.
Soda bottles.

Hogan, Pauline
"A Canadian Legacy in Glass"
Antique Bottle World, 4, No. 4, Sept. 1977, pp. 34–35.
Erie Bottle Glassworks, Ontario.

Hovius, Jim
"Bixler a Pioneer Bottler"
Bottle News, 6, No. 59, Nov. 1977, p. 25, illus.

Hutkowski, Dave
"A List of Connecticut Sodas"
Antique Bottle World, 4, No. 4, Sept. 1977, pp. 30–31.

Johnson, Mary Ellen and Chapman, Bernadine
"H.H.H. Horseman's Hope Horse Medicine: Another California Legend"
Old Bottle Magazine, 2, No. 1, Jan. 1977, pp. 14–16, illus.

Johnson, Mary Ellen and Chapman, Bernadine
"Liquor Bottles in Miniature"
American Collector, 8, No. 5, May 1977, pp. 6–7+, illus.

Kath, Vivian
"Granny Kath's Kitchen"
Continuing series in *Bottle News*, 5, No. 49, Jan. 1977–6, No. 60, Dec. 1977.

Kaukali, Edwin
"That Lost Owl Drug Company of Hawaii"
The Federation Letter, 7, No. 5, May 1977, pp. 11–13, illus.
Medicine bottles.

Knowles, C. S.
Glass and Porcelain Standard Insulators: Catalogue No. 20, 1902 Edition (reprint)
Fresno, Calif.: Frank R. Peters, n.d., 48 pp., illus.

Larned, Larry
"Insulator — Bugs?"
Insulator, 9, No. 10, Dec. 1977, pp. 13–15, illus.
Transcontinental telephone line and problems with insulators.

Lastovica, Al
"Bottle Digging in Cape Town South Africa"
Old Bottle Magazine, 10, No. 12, Dec. 1977, pp. 10–12, illus.
European, particularly British, finds.

Locke Insulator Manufacturing Co.
The Insulator Book; Catalog No. 9, 1907 Edition (reprint)
Fresno, Calif.: Frank R. Peters, n.d., 61 pp., illus.

McCann, Jerry
"Fruit Jar Collecting in America"
Continuing series in *Antique Bottle World*, 3, No. 9, Jan. 1977–4, No. 7, Dec. 1977.

McCurty, Jeff
"A Closer Look at Corning-Pyrex Insulators"
Insulators, 9, No. 10, Dec. 1977, pp. 4–11, illus.

McDOUGALD, CAROL
"Insulator News"
Irregular series in *Antique Bottle World*, 3, No. 9, Jan. 1977–4, No. 7, Dec. 1977.

MASSENGALE, OLIVER
"The Hygeia Nursing Bottle"
Keeping Abreast, 5, No. 2, Dec. 1977, pp. 3–13, illus.

MATTHEWS, GERALD
"Collecting Territorial Hutchinsons"
Old Bottle Magazine, 10, No. 10, Oct. 1977, pp. 40–41, illus.

MATTHEWS, WILLIAM
"Pensacola's Pepsi-Cola"
Old Bottle Magazine, 10, No. 10, Oct. 1977, pp. 10–11, illus.

MAURATH, JOE, JR.
"Insulator By-Lines"
Continuing series in *Old Bottle Magazine*, 10, No. 1, Jan. 1977–10, No. 12, Dec. 1977.

MILHOLLAND, MARION AND EVELYN
More about Glass Insulators
Spanaway, Wash.: Marion C. Milholland, 1976, 455 pp., illus.

MILLER, EVERETT AND ADDIE
"Barth-Art Glass Co."
Rainbow Review, 7, No. 9, Sept. 1977, pp. 28–29, illus.

MONNETTE, JEANNE
"Wonderful World of Avon"
Irregular series in *National Depression Glass Journal*, 5, No. 10, Jan. 1977–6, No. 9, Dec. 1977.

MONTGOMERY, BOB
"Collecting Inks"
Old Bottle Magazine, 10, No. 9, Sept. 1977, pp. 5–7, illus.

MONTGOMERY, BOB
"Mineral and Soda Water Bottle Collecting"
Antique Bottle World, 3, No. 11, March 1977, pp. 5–6.

MUNSEY, CECIL
"Who Was Charley Ross?"
The Federation Letter, 7, No. 8, Sept. 1977, pp. 14–15, illus.
Hair oil bottles.

NAMIAT, ROBERT
Barber Bottles with Prices
Des Moines, Iowa: Wallace-Homestead Book Co., 1977, n.pag., illus.

O'CONNELL, ANNETTE
"The Booz Bottle"
Old Bottle Magazine, 10, No. 7, July 1977, pp. 54–55, illus.

O'CONNELL, ANNETTE
"Prohibition Bottles and Collectibles"
Old Bottle Magazine, 10, No. 9, Sept. 1977, pp. 36–38+, illus.
Part II: 10, No. 10, Oct. 1977, pp. 6–8, illus.

ODELL, JOHN
"Dirtless Digging"
Antique Bottle World, 3, No. 10, Feb. 1977, pp. 24–30, illus.
Ways to research a bottle's age and history.

OWENS-ILLINOIS, INC., comp.
"A Brief History of American Bottles"
American Antiques, 5, No. 8, Aug. 1977, pp. 29–33, illus.

PETERS, FRAN
"Discovering Perfumes"
Irregular series in *Old Bottle Magazine*, 10, No. 1, Jan. 1977–10, No. 12, Dec. 1977.

PFEIFFER, EDDIE
"That Elusive Jenny Lind Flask"
The Heritage Newsletter, 8, No. 3, March 1977, p. 11.

ROBERTUS, BARBARA
"Minnesota's Man and Beast"
The Federation Letter, 7, No. 1, Jan. 1977, pp. 6–9, illus.
Medicine bottles.

ROLLER, DICK
"Fruit Jar News"
Continuing series in *Old Bottle Magazine*, 10, No. 1, Jan. 1977–10, No. 12, Dec. 1977.

RUPPRECHT, LOUIS
Hotel, Bar-Room, Bottlers' Supplies: Catalogue No. 2
Scranton, Pa.: Louis Rupprecht, n.d. [c. 1870's], n.pag., illus.

RZECINSKI, EILEEN
"Contemporaries"
Continuing series in *Antique Bottle World*, 3, No. 9, Jan. 1977–4, No. 7, Dec. 1977.

SELLARI, DOT AND CARLO, eds.
Official Guide to Bottles New and Old
Florence, Ala.: House of Collectibles, 1977, 432 pp., illus.

SIELING, SUSAN
"Frisco Fruit Jar"
Old Bottle Magazine, 10, No. 8, Aug. 1977, p. 17, illus.

SIELING, SUSAN
"Jar Collecting in Australia"
Antique Bottle World, 4, No. 7, Dec. 1977, pp. 10–11, illus.

SOSNOWSKI, KEN
"Borden's—It's the Cream of the Crop"
Antique Bottle World, 4, No. 2, July 1977, pp. 6–8, illus.

SOSNOWSKI, KEN
"Containers That Wear Sunglasses"
Antique Bottle World, 4, No. 3, Aug. 1977, pp. 24–25, illus.
Amber milk bottles.

SPEAR, ANN E.
"G. W. Merchant's Carbonic Mineral Water Mystery Solved by Rare Invoice Find"
Old Bottle Magazine, 10, No. 12, Dec. 1977, pp. 13–14, illus.
Lockport Glass Works.

TANSLEY, JUNE
"Old 'Masters' from the United Kingdom"
Old Bottle Magazine, 10, No. 8, Aug. 1977, pp. 14–15, illus.
British inks.

TOKYO. SUNTORY ART MUSEUM
Bottles of the World (translated title)
Tokyo: Suntory Art Museum, 1977, n.pag., illus.
Exhibit of bottles from Egypt to 19th c.

VANDER ZANDEN, MARIANNE
"Coca-Cola, Then and Now: The Shape of Things to Come"
Antique Bottle World, 4, No. 2, July 1977, pp. 16–17, illus.

VOYTEK, MIKE
"Halloween Auction in Bolton, Massachusetts Sees Spirited Bidding"
Old Bottle Magazine, 2, No. 1, Jan. 1977, pp. 8–13, illus.
Sale of Wiedeman and Rowland collection of bottles.

WEEDEN, CYRIL
"The Glass Container Industry 1916 to 1976"
Glass Technology, 17, No. 5, Oct. 1976, pp. 165–181, illus.

Weitzman, David
Underfoot; An Everyday Guide to Exploring the American Past
New York: Charles Scribner's Sons, 191 pp., illus.
Chapter on American glassmaking, especially bottles.

Welko, Paul
"Did You Know about Chicago Hutchinsons?"
Antique Bottle World, 4, No. 3, Aug. 1977, pp. 14–15, illus.

Whitall Tatum Company; Price List 1912
Philadelphia, Pa.: Whitall Tatum Co., 1912, 58 pp., illus.
Pharmaceutical glassware.

White, Leora
"Inks Uncorked"
The Federation Letter, 7, No. 10, Nov. 1977, pp. 10–12, illus.

Zumwalt, Betty
"Sauce for the Goose"
Antique Bottle World, 3, No. 9, Jan. 1977, pp. 28–31, illus.
Worcestershire sauce bottles.

RUSSIAN

Asarina, N.
"Russkoe narodnoe steklo XVIII veka"
Dekorativnoe Iskusstvo SSSR, Jan 1974, pp. 56–57, illus.
18th c. enameled bottles.

Leningrad. The Russian Museum
Russian Applied Art; Eighteenth to Early Twentieth Century
Leningrad: Aurora Art Publishers, 1976, 199 pp., illus.

Lesley, Parker
Fabergé; A Catalog of the Lillian Thomas Pratt Collection of Russian Imperial Jewels
Richmond, Va.: Virginia Museum, 1976, 160 pp., illus.
Rock crystal, enamels.

London. Victoria and Albert Museum
Fabergé
London: Victoria and Albert Museum, 1977, n.pag., illus.

London. Victoria and Albert Museum
Fabergé 1846–1920
London: Debrett's Peerage Ltd. for the Victoria and Albert Museum, 1977, 132 pp., illus.
Enamels and rock crystal.

Ukrainian Folk Glassware
Leningrad: Aurora Art Publishers, 1975, folder of 16 photos with short text.
From the Ukrainian SSR Museum of Ukrainian Decorative Folk Art, Kiev.

SOUTH AND EAST ASIAN

Cammann, Schuyler
Substance and Symbol in Chinese Toggles
Philadelphia: University of Pennsylvania Press, 1962, 256 pp., illus.

Hitt, Henry
Old Chinese Snuff Bottles
Bremerton, Wash.: Henry C. Hitt, 1945, Snuff Bottle Collector Reprint No. 7, 1977, 110 pp., illus.

Kaynes, Michael
"Containers for the Fragrant Leaf"
Sawasdee (Thai Airways inflight magazine), 5, No. 3, Jan.–Feb. 1977, pp. 10–12, illus.
Snuff bottles.

Legarda, Angelita
"Antique Beads of the Philippine Islands"
Arts of Asia, 7, No. 5, Sept./Oct. 1977, pp. 61–70, illus.

Moss, Hugh
By Imperial Command; An Introduction to Ch'ing Imperial Painted Enamels
Hong Kong: Hibiya Company Ltd., 1976, 2 vols., 98 pp., illus.

Nara, Japan. Nara National Museum
Special Exhibition of the Shōsōin Treasures; Illustrated Catalogue
Nara: Nara National Museum, 1955, n.pag., illus.
Glass game pieces.

New York, New York. China House Gallery
Early Chinese Miniatures (text by Paul Singer)
New York: China Institute in America, 1977, 95 pp., illus.
Mirror for a tomb and a figurine of a lion.

Oriental Snuff Bottles
Palo Alto, Calif.: Mayfield Publishing Co., [1977], 157 pp., illus.
An anonymous collection.

Peking. The Palace Museum
The Selected Handicrafts from the Collections of The Palace Museum
Peking: Wen Wu Press, 1974, n.pag., illus.

Satô, Junshirô
"About the Two-pot Glass Furnaces: an Observation for Analogical Inference on Small-sized Kilns of the Tokugawa Period"
Glass (Garasu), Association for Glass Art Studies, No. 1, Dec. 1975, pp. 17–25, illus., English summary.

Tanahashi-Junji
"About the 'Wa Biidoro Seisaku-hen' and Its 'Appendix'; I"
Glass (Garasu), No. 1, Dec. 1975, pp. 1–8, English summary.
An 1830 manuscript about glass manufacturing.

Tanahashi-Junji
"About the 'Wa Biidoro Seisaku-hen' and Its 'Appendix'; II"
Glass (Garasu), No. 2, Sept. 1976, pp. 4–10, illus., English summary.

Yoshimizu, T.
"The Green Glass Cinerary Urn for Fumi no Nemaro and the Green Glass Cinerary Urn Excavated at Miyajidake Jinja"
Kokogaku Zasshi, 57, No. 2, Nov. 1971, pp. 56–96, illus.

MODERN

ART NOUVEAU AND ART DECO

Anonymous
"Corning Gets 1905 Tiffany Window"
The Collector, 8, No. 311, Feb. 1977, p. 17.
"Tiffany Window Displayed at Corning, N.Y., Museum of Glass"
Hobbies, 82, No. 1, March 1977, p. 118 and cover.
"Trois Livres sur Emile Gallé"
Connaissance des Arts, 275, Jan. 1975, p. 93.

ADAMS, JOHN D.
Lamps and Shades in Metal and Art Glass
Chicago: Popular Mechanics Co., 1911, 114 pp., illus.

BAKER, RICHARD
"The Accrington Tiffany Find"
The Australasian Antique Collector, 17, 1976, pp. 85–88, illus.
Favrile glass in the Haworth Art Gallery collection.

BRUNHAMMER, YVONNE
Le Style 1925
Paris: Baschet et Cie, [1977], 191 pp., illus.
Glass, pp. 156–165.

BRUNHAMMER, YVONNE
1925
Paris: Les Presses de la Connaissance, 1976, 231 pp., illus.
Paris Exposition of 1925.

BUTLER, JOSEPH
"Collecting Tiffany"
The Glass Workshop, 7, No. 3, Winter 1977, pp. 30–32, illus.

CHELMOW, DORIS
"One Man's Collection: Tiffany Glass"
Americana, 5, No. 1, March 1977, pp. 39–43, illus.
Egon Neustadt Collection.

DOROS, PAUL
"Tiffany Glass at the Chrysler Museum, Norfolk, Virginia"
Antiques, 111, No. 4, April 1977, pp. 746–750, illus.

ESCHMANN, KARL
Jugendstil; Ursprünge, Parallelen, Folgen
Saarbrücken: Universitäts- und Schulbuchverlag; Kastellaun; Aloys Henn, 1976, 188 pp., illus.

FRUEH, ERNE AND FLORENCE
"Schneider's Art Deco Glass"
The Antiques Journal, 32, No. 2, Feb. 1977, pp. 18–20, illus.

GARNER, PHILIPPE
"Emile Gallé; Master in Glass, Wood and Faience"
The Antique Dealer & Collectors Guide, Feb. 1977, pp. 70–74, illus.

GRANT, MARENA
"Treasures from Laurelton Hall"
Antiques, 111, No. 4, April 1977, pp. 752–759, illus.

HASLAM, MALCOLM
Marks and Monograms of the Modern Movement 1875–1930
New York: Charles Scribner's Sons, 1977, 192 pp., illus.
Various European, Russian, and American glass marks.

HAYOT, MONELLE
"L'atelier de René Lalique; tradition et création"
L'Oeil, No. 260, March 1977, pp. 22–29, illus.

KOCH, ROBERT
Louis C. Tiffany und seine Glaskunst
Bern: Paul Haupt, 1976, 52 pp., illus.

KOCH, ROBERT
Louis C. Tiffany's Art Glass
New York: Crown Publishers, Inc., 1977, 134 pp., illus.

LALIQUE, MARC AND MARIE-CLAUDE, comp.
Lalique par Lalique
Paris: Société Lalique, 1977, 309 pp., illus.

LEDOUX, THÉRÈSE
"Les Cristalleries du Val-Saint-Lambert ont 150 ans"
Actualités des Arts, 2, No. 12, Oct. 1974, pp. 4–8, illus.

LITTLETON, HARVEY
"Gallé: Transcendence in Glass and Wood"
Craft Horizons, 37, No. 4, Aug. 1977, pp. 32–34, illus.

LONDON. FISCHER FINE ART LTD.
Josef Hoffmann 1870–1956
London: Fischer Fine Art Ltd., 1977, 48 pp., illus.
Tableware, stained glass.

MCCLINTON, KATHERINE MORRISON
"Gallé Glass in Review"
Collector Editions, 5, No. 4, Fall 1977, pp. 45–46, illus.

MEHLMAN, ROBERT
"Antiques: Collecting Lalique Glass"
Residential Interiors, 2, No. 3, May/June 1977, p. 14, illus.

MEHLMAN, ROBERT
"The Glass of René Lalique"
Residential Interiors, 2, No. 2, March/April 1977, p. 12, illus.

NANCY. MUSÉE DES BEAUX-ARTS
Daum; Cent ans de verre et de cristal
Nancy: l'Office de Tourisme de Nancy, 1977, n.pag., illus.

NYSTRÖM, BENGT
"Jugendglas à la Wallander och Wennerberg"
Fataburen, 1975, pp. 47–66, illus., English summary.
Art Nouveau glass by Alf Wallander and Gunnar Wennerberg.

PERCY, CHRISTOPHER V.
Lalique Verrier
Lausanne: Edita Denoël, 1977, 190 pp., illus.

SMITH, ALAN
"Tiffany 'Favrile' Glass at Accrington"
Museums Journal, 76, No. 4, March 1977, pp. 155–156, illus.

THORP, GREGORY
"Cincinnati Union Terminal's Mosaic Panorama"
American Art Review, 4, No. 2, Aug. 1977, pp. 48–65, illus.
Mosaics, 1930's.

Tiffany Studios
New York: Tiffany Studios, n.d., n.pag., illus.
Brief catalog of decorative objects, produced probably in the 1930's.

WADDELL, ROBERTA, ed.
The Art Nouveau Style
New York: Dover Publications, 1977, 288 pp., illus.

WEISS, PEG
"The Judson Memorial Window by Tiffany"
Everson Museum of Art Bulletin, Jan. 1977, n.pag., illus.

CONTEMPORARY (post-1945)

ANONYMOUS
"Artist Creates Lifelike Animals, Flowers, from Kimble Division Glass Rod and Tubing"
Owens-Illinois Outlook, No. 200, Feb. 1977, pp. 1–2, illus.
Glass sculpture by Hans Frabel.

"Coburger Glaspreis '77"
Kunst und Handwerk, 8, Aug. 1977, pp. 276–282, illus.
Coburg exhibition of contemporary European glass.

"Crafts in Brief: Peter Dreiser, Glass Engraver"
Crafts, 26, May/June 1977, p. 9, illus.
German-born engraver now in England.

"Czech Glass: An Ever Open Door"
Tableware International, 7, No. 7, June 1977, pp. 42–43, illus.

"Dominick Labino Receives His 1977 Phoenix Award"
The Glass Industry, 58, No. 11, Nov. 1977, pp. 24–25, illus.
Glass inventor, technologist, artist.

"Ebenbürtige Sonderschau"
Kunst und Handwerk, 20, March 3, 1977, p. 99, illus.

"Glas vor der Lampe; Ausstellung und Fachseminar in Wien"
Kunst und Handwerk, 20, May 5, 1977, pp. 196–197, illus.

"Glas vor der Lampe: ein Symposion"
Kunst und Handwerk, 20, June/July, 1977, pp. 253–254, illus.
Baden-bei-Wien lampworking symposium.

"Glasmuseum im Spessart"
Kunst und Handwerk, 20, No. 10, Oct. 1976, pp. 367–369, illus.
New museum in area traditionally known for glassmaking.

"Glassware and Glass Manufacturing Equipment in Belgium"
Glass, 53, No. 4, 1976, pp. 109–118, illus.

"Harriet Hyams: Bridge Gallery, White Plains, N.Y."
Arts Magazine, 50, No. 10, June 1976, pp. 22–23, illus.

"He Sculptures in Flat Glass"
American Glass Review, 97, No. 11, May 1977, p. 10, illus.
"Spanish glass artist Torres Estaban.

"I vetri felici di Ettore Sottsass"
Casa Vogue, 76, Nov. 1977, pp. 196–197, illus.
Venetian glass designer Ettore Sottsass.

"Karl Dedy, Künstler und Unternehmer"
Glaswelt, 29, No. 10, Oct. 1976, p. 713, illus.
German glass artist.

"A Living Look at Glass"
Pilkington News, May 4, 1977, pp. 6–7, illus.
Pilkington Glass Museum.

"Mettiamosu casa insieme; prodotti per abitare meglio"
Abitare, Supplement 5, 13, No. 160, Dec. 1977, 37 pp., illus.
Lighting devices.

"1977 Phoenix Award Dinner"
The Glass Industry, 58, No. 7, July 1977, pp. 12–14; p. 34, illus.
Dominick Labino, glass technologist and artist-craftsman.

"Rosenthal macht Glas seit 25 Jahren"
Die Schaulade, 52, No. 6, June 1977, pp. 926–933, illus.
Designers at Rosenthal with some of their best work.

"Schaulade Porträts des Monats; Per Lütken 60 Jahre alt."
Die Schaulade, 2, Feb. 1977, p. 397, illus.
Danish glass artist Per Lütken.

"Sklo Jana Frydrycha"
Umění a Řemesla, 1, 1977, p. 9, illus.
Czech glass artist Jan Frydrych.

"Special Exhibit of Corning Glass to Run through October"
The Collector, 8, No. 314, May 1977, p. 6, illus.

"To Visit: Wheaton Village, Millville, New Jersey"
Spinning Wheel, 33, No. 8, Oct. 1977, pp. 28–30, illus.

"Věra Lišková's Glass Sculptures"
Glass Review, 32, No. 7, 1977, pp. 10–14, illus.

"Webb Corbett Crystal"
Tableware International, 6, No. 10, Sept. 1976, p. 29.

ADLEROVÁ, ALENA
"The Death of Karel Hetteš"
Glass Review, 31, No. 12, 1976, p. 17, illus.
Glass historian and critic.

ALLISON, GRACE
"Mayflower Glass Works"
Rainbow Review, 6, No. 12, Dec. 1976, pp. 30–32, illus.

AMES, IOWA.
OCTAGON ART CENTER
Technology and the Artist-Craftsman
Ames, Iowa: Iowa State University, 1973, n.pag., illus.

BAMFORD, JOAN
"Collectable Jubilee Souvenirs"
The Antique Dealer & Collectors Guide, April 1977, pp. 79–82, illus.

BEARD, GEOFFREY
International Modern Glass
London: Barrie & Jenkins, 1976, 264 pp., illus.

BEER, EILEEN HARRISON
Scandinavian Design; Objects of a Life Style
New York: The American-Scandinavian Society, 1975, 214 pp., illus.
Glass pp. 86–123; mostly Swedish designers and firms.

Bohemian Crystal Chandeliers
Liberec, Czechoslovakia: Glassexport, n.d., 44 pp., illus.

BRINDISI. ISTITUTO MAGISTRALE STATALE
Sculture in vetro; E. Costantini
Brindisi: Rotary Club, 1976, n.pag., illus.
Venetian glassmaker.

BRNO. MORAVSKÁ GALERIE
Škrdlovická sklárna a její výtvarníci
Brno: Moravská Galerie, 1977, 48 pp., illus., English summary.
Czech glassworks at Škrdlovice.

BRNO. MORAVSKÁ GALERIE
Svatopluk Kasalý
Brno: Moravská Galerie v Brně, 1972, n.pag., illus.
Glass artist and designer.

BUECHNER, THOMAS
"Coburger Glaspreis 1977"
Craft Horizons, 37, No. 6, Dec. 1977, pp. 16–27, illus.

COBURG. KUNSTSAMMLUNGEN DER VESTE COBURG
Coburger Glaspreis 1977 für moderne Glasgestaltung in Europa
Coburg: Heino Maedebach, 1977, 340 pp., illus., English text
Major exhibition of contemporary European work.

COBURG. KUNSTSAMMLUNGEN DER VESTE COBURG
Kyohei Fujita
Coburg: Kunstsammlungen der Veste Coburg, 1977, n.pag., illus.
Japanese glass designer.

COLUMBUS, OHIO. COLUMBUS GALLERY OF FINE ARTS
Beaux Arts Designer/Craftsman '77
Columbus: Beaux Arts, affiliate of Columbus Gallery of Fine Arts, 1977, n.pag., illus.
Midwestern glass artists.

CRAIG, ROSE MARY
"The Artisans of Ireland: Crafts as Collectibles"
Acquire, 5, No. 1, March 1977, pp. 9–15, illus.
Glass designer Simon Pearce.

CRISTALLERIE DAUM
Sculptures en pâte de verre
Nancy: Cristallerie Daum, 1977, 16 pp., illus.

CURRY, MIMI AND LICHTMAN, LINDA
"Pilchuk: Two Views"
Glass (formerly *Glass Art*), 5, No. 1, Jan. 1977, pp. 15–19, illus.

DANIELIS, JAN
"Care for the Development of Industrial Design and the Artistic Standard of Products in the Glass Industry"
Glass Review, 32, No. 5, 1977, pp. 9–23, illus.

DE KOONING, ELAINE
"Jeanne Reynal: Mosaic Portraits"
Craft Horizons, 36, No. 4, Aug. 1976, pp. 33–35, illus.

DOROS, PAUL
"Contemporary Art Glass '76 Exhibit"
Hobbies, 82, No. 1, March 1977, pp. 34CC–34FF, illus.

DRESDEN. MUSEUM FÜR KUNSTHANDWERK DRESDEN
Friedrich Bundtzen; 20 Jahre Glasgestaltung in der DDR
Dresden: Staatlichen Kunstsammlungen, 1969, 59 pp., illus.
Tableware designer.

DULUTH, MINNESOTA. TWEED MUSEUM OF ART
Lake Superior 77
Duluth, Minn.: University of Minnesota, 1977, n.pag., illus.
Numerous glass entries.

DÜSSELDORF. KUNSTMUSEUM.
Benny Motzfeldt: Glas aus Norwegen
Münster: Deutsch-Norwegische Gesellschaft, 1977, 37 pp., illus.

ECKEL, LUCILE
"Robert Ray Banford, Paperweight Artisan"
Annual Bulletin of the Paperweight Collectors' Association, 1977, pp. 17–19, illus.

EL PASO, TEXAS. EL PASO MUSEUM OF ART
The El Paso Designer Craftsmen
El Paso: El Paso Museum of Art, 1977, 12 pp., illus.

ERFURT. INTERNATIONALE GARTENBAUAUSSTELLUNG
I. Quadriennale des Kunsthandwerks sozialistischer Länder; Erfurt 1974; Dokumentation
Erfurt: Kunsthandwerk sozialistischer Länder, 1974, 308 pp., illus.
Glass sculpture and tableware.

ESSEN. GALERIE SCHAUMANN
Stefan
Essen: Galerie Schaumann, [1977], 11 pp., illus.
Exhibit of Stefan von Reiswitz' glass/plexiglass work.

ESSEN. GALERIE SCHAUMANN
Stefan
Málaga, Spain: Gráficas Sorima, [1976], n.pag., illus.

FRANZENSBAD BADEN B. WIEN. GLASSTUDIO
Glas vor der Lampe
Franzensbad Baden B. Wien: Glasstudio, 1977, n.pag., illus.
Exhibit and seminar on lampworking.

GETTY, NILDA FERNANDEZ
Contemporary Crafts of the Americas: 1975
Chicago: Henry Regnery Co., 1975, 172 pp., illus.
A few examples of U.S. glass.

GEVELSBERG. RATHAUS-FOYER
Hans Peter Kremer
Gevelsberg: [H. P. Kremer?], 1976, n.pag., illus.
German glass designer and sculptor.

GLUCK, JAY AND GLUCK, SUMI, H., eds.
A Survey of Persian Handicraft
Teheran: Bank Melli Iran, 1977, 416 pp., illus.

GOLDSCHMIT, WERNER
"Hanns Model der Glasschnitzer"
Kunst und Handwerk, 20, March 3, 1977, pp. 108–111, illus.
Glass cutting artist.

GREENSBURG, PA. THE WESTMORELAND COUNTY MUSEUM OF ART
Dudchenko; Zandhuis; Dudchenko
Greensburg: The Westmoreland County Museum of Art, 1976, n.pag., illus.
Artists Jan Zandhuis, Boris and Nancy Dudchenko.

GUEDES, P.
"Wall of Glass Faces"
RIBA (Royal Institute of British Architects) *Journal*, 83, Oct. 1977, p. 432, illus.
Design for a Venetian fountain.

'Gusebskii Khrustal'
Moscow: Vladimiro Suzdal'skii Istoriko Khudozhestvennyi i Arkhitekturnyi Musei-Zapovednik, 1974, n.pag., illus.
Folder of tableware.

HAVEL, ROBERT
"Josefodol — An Important Work of the Bohemia Glassworks at Poděbrady"
Glass Review, 31, No. 12, 1976, pp. 7–11, illus.

HEJTMAN, JOSEF
"Large Lighting Fixtures from Kamenický Šenov"
Glass Review, 32, No. 7, 1977, pp. 20–24, illus.

HOFFMAN, RUDOLF
Lauschaer Glaskunst heute
Lauscha: Museum für Glaskunst Lauscha and VEB Glaskunst Lauscha, 1972, 95 pp., illus.

HOLLISTER, PAUL
"Discovering What Corning Collects"
Acquire, 5, No. 2, May 1977, pp. 22–23, illus.

JACOBSEN, ROGER AND DRABECK, BERNARD
"Perthshire Silhouettes"
Annual Bulletin of the Paperweight Collectors' Association, 1977, pp. 33–39, illus.
Contemporary paperweights.

KÄMPFER, FRITZ
"Glaskunst in der DDR"
Silikattechnik, 28, No. 6, June 1977, pp. 187–190, illus.

KASSEL. HESSISCHES LANDESMUSEUM
Glaskunst der Gegenwart
Kassel: Staatliche Kunstsammlungen, 1977, 40 pp., illus.

KIPLING, KATHRYN
"Glassblower Jean Dove"
Rainbow Review, 7, No. 6, June 1977, pp. 34–35, illus.

KUTNER, JANET
"Larry Bell's Iceberg"
Arts Magazine, 50, No. 5, Jan. 1976, pp. 62–66, illus.
Glass sculpture-environment.

LANGHAMER, ANTONÍN
"Deliberations on the Results of the Co-operation between Ludvika Smrčková and Adolf Matura"
Glass Review, 31, No. 12, 1976, pp. 12–16, illus.
"Engraved Glass of the 20th Century — Part I; The School of Professor Joseph Drahoňovský"
Glass Review, 30, No. 9, 1975, pp. 5–12, illus.
"An Exhibition of Glass at Železný Brod"
Glass Review, 31, No. 12, 1976, pp. 26–28, illus.
"Glass Artist Felix Průša"
Glass Review, 32, No. 7, 1977, pp. 5–7, illus.
"Jan Štibych's Window Glass"
Glass Review, 31, No. 10, 1976, pp. 10–14, illus.

LEIPZIG. MUSEUM DES KUNSTHANDWERKS (GRASSIMUSEUM)
Glaskunst in der DDR
Leipzig: Museum des Kunsthandwerks, 1977, n.pag., illus.
15 contemporary glass artists of East Germany.

LEIPZIG. MUSEUM DES KUNSTHANDWERKS (GRASSIMUSEUM)
Ulrike und Thomas Oelzner, Glas; Rudolf Oelzner, Plastik
Leipzig: Museum des Kunsthandwerks, 1977, 32 pp., illus.

Leningradskii zavod khudozhestvennogo stekla (Leningrad Works of Artistic Glassware)
Leningrad: [Leningrad Art Glass Works], n.d., [1976?], n.pag.
Small illustrated folder.

Liberty Village
Flemington, N.J.: Liberty Village Foundation, [1977], n.pag., illus.
Glassmaking at a recreated colonial village.

LOO, BERT VAN
"The London Conference—Art and Industry Must Merge"
Glass (formerly *Glass Art*), 5, No. 6, Oct. 1977, pp. 55–56, illus.
Hot glass conference.

MARŠÍKOVÁ, J.
"Glass from Vrbno pod Pradědem"
Glass Review, 31, No. 12, 1976, pp. 22–25, illus.

MARŠÍKOVÁ, J.
"The Glassworks at Libochovice"
Glass Review, 31, No. 10, 1976, pp. 2–6, illus.

MARŠÍKOVÁ, J.
"Models from Nature"
Glass Review, 31, No. 10, 1976, pp. 24–25, illus.
Glass vegetables by František Koudelka.

MARTILIK, JAROSLAV
"Die Lage der Glas- und keramischen Industrie in der tschechoslowakischen Volkswirtschaft"
Silikattechnik, 27, No. 6, 1976, pp. 183–186, illus.

MYERS, JOEL PHILIP
"New American Glass"
Craft Horizons, 36, No. 4, Aug. 1976, pp. 36–41, illus.

MOYNEHAN, BARBARA
"Glass Art Revived in Studio Movement"
The New York–Pennsylvania Collector, 1, No. 6, June 1976, pp. 2+, illus.

NEW YORK. GALERIE INTERNATIONALE
Torres Esteban
New York: Galerie Internationale, 1976, n.pag., illus.
Spanish glass sculptor.

PAVLISTA, RUDOLF
"Glasexpansion in der Tschechoslowakei"
Die Schaulade, 12, Dec. 1976, pp. 1744–1745+, illus.

PENFIELD, LOUIS
"Ingeborg Lundin of Orrefors"
American Scandinavian Review, 57, No. 1, March 1969, pp. 44–49, illus.

PERRY, ANTHONY
"New Paul Joseph Stankard Orchids"
Annual Bulletin of the Paperweight Collectors' Association, 1977, pp. 23–26, illus.

PIENE, OTTO
"Glas, Gas und Elektrizität"
Du, 37, Aug. 1977, pp. 62–74, illus.
Contemporary studio glass artists.

PRAGUE. GALERIE NA BETLÉMSKÉM NÁMĚSTÍ
Sklo severočeských výtvarníků (text by Alena Adlerová)
Prague: Galerie na Betlémském náměstí, 1969, n.pag., illus.
Work of 30 contemporary glass artists.

PRAGUE. GALERIE VÁCLAVA ŠPÁLY
Umělecké sklo 1977
Prague: Svaz Českých Výtvarných Umělců, 1977, n.pag.
Exhibit of about 50 contemporary Czech glass artists. No illus.

PRAGUE. VÝSTAVNÍ SÍŇ FRONTA
František Vízner—Sklo 1961–1971
Prague: Výstavní Síň Fronta, 1971, n.pag., illus.

RATH, PETER, ed.
"The Glass Workshop at the VII World Crafts Congress in Mexico"
Glass (formerly *Glass Art*), 5, No. 3, April 1977, pp. 25–48, illus.

ROGERS, JO ANN
"Ric Iorio and His Art Glass"
The Antiques Journal, 32, No. 4, April 1977, pp. 22–24, illus.

ROSTOCK. KUNSTHALLE
Glaskunst aus Lauscha
n.l. (Dessau?): Zentrum für Kunstausstellungen der DDR, 1976, 20 pp., illus.
Work of seven Lauscha glass artists.

ROVERA, MARINA
"E'giovane lo spirito di Orrefors"
Casa Vogue, No. 75, Oct. 1977, pp. 192–193, illus.

RUDOI, B. L.
Novaia zhizn' stekla (Glass for Modern Living)
Moscow: Izdatel'stvo "Znanie," 1963, 47 pp.

SACRAMENTO, CALIFORNIA. E. B. CROCKER ART GALLERY
California Crafts X
Sacramento: The Creative Arts League of Sacramento, 1977, 79 pp., illus.
Includes six glass artists.

SACRAMENTO, CALIFORNIA. E. B. CROCKER ART GALLERY
The Ninth Biennial Crafts Exhibition
Sacramento: The Creative Arts League of Sacramento, 1975, 78 pp., illus.
Twelve glass artists.

SCHÄFER, MANFRED
"Lausitzer Glas"
Silikattechnik, 27, No. 2, 1976, pp. 52–54, illus.
Lusation glass.

Sherengovaja, Z.
"Eksperimenty khudoznikov i tekhnologov"
Dekorativnoe Iskusstvo SSSR, Feb. 1974, pp. 32–36, illus.
Contemporary Russian tablewares.

Sisson, Barbara
"John Gentile, Paperweight Artist; Part One"
Rainbow Review, 7, No. 1, Jan. 1977, pp. 26–29, illus.

Sisson, Barbara
"John Gentile, Paperweight Artist; Part Two"
Rainbow Review, 7, No. 2, Feb. 1977, pp. 26–29, illus.

Souffler le verre à Dieulefit; Claude et Florence Morin
Lausanne: Imprimerie Bron Sa, n.d., n.pag., illus., English summary.
French glassmaking couple.

Tallin, Estonia.
Gorodskoy Museum
Steklo Leidy Iurgen
Tallin: Tallinskiĭ Gorodskoí Muzei, 1976, 62 pp., illus.
Glass artist Leidy Iurgen.

Tilche, Paolo
"Il vetro"
Abitare, 13, No. 160, Dec. 1977, pp. 30–41, illus.
Tableware.

Vamberecká, J.
"Jiri Repásek's Lead Crystal"
Glass Review, 32, No. 4, 1977, pp. 18–21, illus.

Vamberecká, J.
"Off-hand Shaped Glass From Chribská"
Glass Review, 32, No. 6, 1977, pp. 26–27, illus.

Vamberecká, J.
"Pressed Lead Crystal"
Glass Review, 31, No. 10, 1976, pp. 18–23, illus.

Vojta, J. M.
"Jana Míčková's Individual Lighting Fixtures"
Glass Review, 32, No. 6, 1977, pp. 13–15, illus.

Vojta, J. M.
"The New Lenora Glassworks"
Glass Review, 31, No. 12, 1976, pp. 18–21, illus.

Vojta, J. M.
"Rosice and Pressed Glass"
Glass Review, 32, No. 6, 1977, pp. 20–23, illus.

Welton, Peter
"Crafts in Question; Whitworth Art Gallery, Manchester"
Crafts, 26, May/June 1977, pp. 46–47, illus.
Pauline Solven's glass.

Wickman, Kerstin
"'Friheten är som att stå framför ett stup'" (Freedom Is Like Standing before an Abyss)
Form, 73, No. 1, 1977, pp. 36–39, illus., English summary.
Swedish glass designer, Bertil Vallien.

Wykes-Joyce, Max
"Glass and the Modern Artist"
The Antique Dealer & Collectors Guide, April 1977, pp. 100–102, illus.
Some European engravers and designers.

STAINED GLASS

Anonymous
"Cathedral Windows Shown at Steuben"
The Gaffer, 35, No. 3, April 1977, p. 7, illus.
"Glass: Beveled and Colored"
Interior Design, 48, No. 4, April 1977, pp. 188–191, illus.
Georgia artisan David Clymer.
"Stained Glass Business Meets Growing Demand"
American Glass Review, 97, No. 8, Feb. 1977, pp. 12–13, illus.
"The 'Vasari' Diary: A Devotion to Stained Glass"
Art News, 76, No. 5, May 1977, pp. 20–22, illus.
Abraham Rattner's window for the Chicago Loop Synagogue.

Abrams, Fred
"Caroling's Wholeo"
Glass (formerly *Glass Art*), 5, No. 3, April 1977, pp. 13–14, illus.
A stained glass dome done by artist Caroling.
"The Fresno Arts Center, California Contemporary Stained Glass Exhibition"
Glass (formerly *Glass Art*), 5, No. 2, Feb. 1977, pp. 20–28+, illus.
"Interviews: Art in Architecture"
Glass (formerly *Glass Art*), 5, No. 6, Oct. 1977, pp. 38–45, illus.
Robert Marquis, Kathy Simon.
"Living through Fresno: An Interview with Dan Fenton"
Glass (formerly *Glass Art*), 5, No. 2, Feb. 1977, pp. 29–45, illus.
"Stained Glass Education: An Introduction"
Glass (formerly *Glass Art*), 5, No. 1, Jan. 1977, pp. 11–12, illus.

Alcaide, Víctor Nieto
La vidriera del renacimiento en España
Madrid: Instituto Diego Velázquez, del Consejo Superior de Investigaciones Científicas, 1970, 63 pp., illus.

Anderes, Bernhard
"Zur Freiburger Glasmalerei des 16. Jahrhunderts"
Unsere Kunstdenkmäler, 26, 1975, pp. 285–290, illus.

Bangs, Jeremy
"Maerten van Heemskerck's 'Bel and the Dragon' and Iconoclasm"
Renaissance Quarterly, 30, No. 1, Spring 1977, pp. 8–11, illus.

Barnett, S., Weiss, D., and Bunnell, K.
"How I Learned: Three Artists Talk"
Glass (formerly *Glass Art*), 5, No. 1, Jan. 1977, pp. 23–25, illus.

Barrié, Roger
"Les verres gravés et l'art du vitrail au XVIe siècle en Bretagne occidentale"
Annales de Bretagne Pays Oest, 83, No. 1, 1976, pp. 35–44.

Bartos, Marshall
"The Stained Glass Studio"
Glass (formerly *Glass Art*), 5, No. 5, June 1977, pp. 8–9+, illus.
Columbia Art Glass Co., Milwaukee.

Baudot, M. and Lafond, J.
Églises et vitraux de la région de Pont-Audemer
n.l.: Nouvelles de l'Eure, n.d., 68 pp., illus.

Becker, Theo
"Making a Windowless Restaurant Big and Beautiful As All Outdoors"
Stained Glass, 72, No. 2, Summer 1977, pp. 89–91, illus.

BECKSMANN, RÜDIGER
"Glasmalereien"
Die Weltkunst, 47, No. 9, May 1, 1977, pp. 910–911, illus.

BERNSTEIN, PAUL
"Roger Darricarrere"
Glass (formerly *Glass Art*), 5, No. 3, April 1977, pp. 19–24+, illus.
Stained glass artist.

BERNSTEIN, PAUL
"Stained Glass Education: It's Better Than Sitting in Front of the Tube"
Glass (formerly *Glass Art*), 5, No. 1, Jan. 1977, pp. 12–14+, illus.

BEYER, V., CHOUX, J., AND LEDEUR, L.
Vitraux de France du Moyen-Âge à la Renaissance; Alsace, Lorraine, Franch-Compté
Colmar-Ingersheim: Editions SAEP, 1970, 256 pp., illus.

Booth Art Glass Co., Special Catalogue No. 769
Buffalo, New York: Booth Art Glass Co., 1909, 64 pp., illus.
Ornamental and stained glass.

COLEMAN, MARIGOLD
"Glass from the North"
Crafts, 26, May/June 1977, pp. 17–21, illus.
Brian Clarke, English architectural glass artist.

CULLENBINE, CAROLE
Stained Glass Reflections
Palo Alto, California: Hidden House, 1976, 64 pp., illus.
How-to-do-it pattern book, mirrors.

DAHOOD, THOMAS
"Stained Glass Design"
Design, 78, No. 3, Midwinter 1977, pp. 20–22, illus.

DOWNS, ARTHUR C.
"Stained Glass in American Architecture"
Nineteenth Century, 3, No. 4, Winter 1977, pp. 54–60, illus.

DRACHENBERG, E., MAERCKER, K-J., AND SCHMIDT, C.
Die mittelalterliche Glasmalerei in Erfurt; Teil 1: Ordenskirchen und Angermuseum
Vienna: Hermann Böhlaus, 1976, 278 pp., illus.

EDWARDS, MARGARET
The Studios of Lancaster's Artists in Stained Glass
Lancaster: Lancaster City Museum, 1974, 10 pp.

THE FLANAGAN & BIEDENWEG COMPANY
Official Catalog. The National Ornamental Glass Manufacturers Association. Special Catalog No. 462.
Chicago, Ill.: The National Ornamental Glass Manufacturers Association, 1909, 64 pp., illus.

FOURNÉE, JEAN
Le jugement dernier d'après le vitrail de la Cathédrale de Coutances
Paris: Jean Fournée, 1964, 175 pp., illus.

FRANZKE, JOHN
"Quarries and Chromatic Structuralism"
Glass (formerly *Glass Art*), 5, No. 6, Oct. 1977, pp. 15–17, illus.

FRAVEL, JANE
"Tessellations . . . How They Were Conceived"
Stained Glass, 72, No. 3, Fall 1977, pp. 171–173, illus.

FROIDEVAUX, YVES-MARIE
"Le vitrail dans l'architecture médiévale"
Les Monuments Historiques de la France, 1, 1977, pp. 31–51, illus.

GALLIN, MARTIN
"Life with a Stained Glass Junkie"
Stained Glass, 72, No. 1, Spring 1977, pp. 48–50, illus.
Artist Saara Gallin.

GAMLEN, ST. J. O.
Ancient Window Glass at Chicksands Priory, Bedfordshire
London: John Roberts Press, 1974, 28 pp., illus.

GIBBS, SUNNY
"Glass and Light, A Multifaceted Medium"
Los Angeles Times Home Magazine, April 10, 1977, pp. 10–11, illus.

GICK, JAMES
Patterns for Stained Glass
Laguna Hills, California: Future Crafts Today, 1977, 48 pp., illus.

GRODECKI, LOUIS
"Esthétique ancienne et moderne du vitrail roman"
Les Monuments Historiques de la France, 1, 1977, pp. 17–30, illus.
Chartres Cathedral.

"La salle du trésor de la cathédrale de Châlons-sur-Marne; les vitraux"
Les Monuments Historiques de la France, 4, Oct.–Dec., 1957, pp. 186–190, illus.

Le Vitrail roman
Fribourg: Office du Livre, 1977, 308 pp., illus.

HARRIS, ESTHER AND BOWMAN, DORIS
"The American Indian in Stained Glass"
Stained Glass, 72, No. 1, Spring 1977, pp. 20–27, illus.

HARTMANN, GEORG
The Goetheanum Glass-Windows
Dornach, Switzerland: Philosophisch-Anthroposophischer Verlag, 1972, 75 pp., illus.

HAUG, JANS, and others
La cathédrale de Strasbourg
Strasbourg: Editions des Dernières Nouvelles, 1957, 149 pp., illus.

HAYWARD, JANE
"Canterbury Comes to New York"
Glass (formerly *Glass Art*), 5, No. 5, June 1977, pp. 16–18+, illus.

HERMAN, STEWART
"Marine Mosaic . . . The Lost Art of Walter Cole Brigham"
Stained Glass, 72, No. 1, Spring 1977, pp. 31–35, illus.

JOHNSON, EDWARD P.
The Windows of the First Congregational Church. [Toledo, Ohio]
Toledo, Ohio: typescript paper of talk by Edward Johnson, 1973, 7 pp.
Tiffany stained glass windows.

KING, DAVID J.
Stained Glass Tours around Norfolk Churches
Norfolk: The Norfolk Society, 1974, 40 pp., illus.

LABRET, JEAN-SIMON
Le vitrail, premières notions
Paris: Dessain et Tolra, 1975, 61 pp., illus.

LANDOLT, ELISABETH
"Die Fenster- und Wappenschenkungen des Standes Basel, 1556–1626"
Zeitschrift für Schweizerische Archäologie und Kunstgeschichte, 34, No. 2, 1977, pp. 113–136, illus.

LAUSANNE. MUSÉE HISTORIQUE DE L'ANCIEN-ÉVÊCHÉ
Cathédrale de Lausanne: 700e anniversaire de la consécration solennelle; Catalogue de l'Exposition
Lausanne: Musée Historique de l'Ancien-Évêché, 1975, 222 pp., illus.
Stained glass windows on life of St. John the Baptist.

LEE, LAWRENCE
The Appreciation of Stained Glass
London: Oxford University Press, 1977, 118 pp., illus.

LEWIS, NANCY
"George Sell: Working in Leaded Glass"
Stained Glass, 72, No. 4, Winter 1977–78, pp. 236–239, illus.

LEYMARIE, JEAN
Marc Chagall; The Jerusalem Windows
New York: George Braziller, Inc., 1962, 210 pp., illus.

LILLICH, MEREDITH
"A Stained Glass Apostle from Sées Cathedral (Normandy) in the Victoria and Albert Museum"
The Burlington Magazine, 119, No. 892, July 1977, pp. 497–500, illus.

LILLICH, MEREDITH
"Les vitraux de Saint-Pierre de Chartres"
Les Monuments Historiques de la France, 1, 1977, pp. 52–57, illus.

LONDON.
THE NATIONAL GALLERY
Late Gothic Art from Cologne
London: The National Gallery, 1977, 164 pp., illus.

LORIN, CH.
Les vitraux du Moyen Âge, les vitraux de la Renaissance
Chartres: Société Archéologique d'Eure-et-Loir, 1909, 32 pp.

MAES, P. V.
"Nicolaas Ruterius en de brandlassuite met de geschiedenis van Saint-Nicolaas"
Arca Lovaniensis, 2, 1973, pp. 181–208, illus., French summary.

MAES, P. V.
"Oud Leuvens brandglas in Engeland"
Arca Lovaniensis, 1, 1972, pp. 189–201, illus., French summary.
Panels and medallions by 16th–17th c. Louvain painters.

MANN, PAUL
"Ten Windows Tell the History of the City"
Stained Glass, 72, No. 1, Spring 1977, pp. 36–38, illus.
Kansas City's City Hall.

MARKS, RICHARD
"Medieval Stained Glass in Bedfordshire: I"
Bedfordshire Magazine, 15, No. 117, Summer 1976, pp. 179–184, illus.

MARKS, RICHARD
"Medieval Stained Glass in Bedfordshire: II"
Bedfordshire Magazine, 15, No. 118, 1976, pp. 228–233, illus.

MARSAT, ANDRE
Cathédrale de Troyes: les vitraux
Paris: C. C. P. Tetraktys, n.d., 108 pp., illus.

MILLARD, RICHARD
"Stained Glass Renaissance Is Moving out of the Churches and into the Secular Sphere"
American Glass Review, 98, No. 1, July 1977, pp. 38–39, illus.

MYERS, MARY
"Stained Glass on the Third Mesa"
Stained Glass, 72, No. 1, Spring 1977, pp. 28–30, illus.

NEWMAN, THELMA, NEWMAN, J., AND NEWMAN, L.
The Lamp and Lighting Book
New York: Crown Publishers, Inc., 1976, 246 pp., illus.
How-to-do-it book, utilizing stained glass.

NEWTON, R. G.
"Tracing the Origins and Development of Stained Glass"
Glass, 53, No. 12, 1976, pp. 390–396, illus.

NIELSEN, BODIL
"Renaissance in Stained Glass"
Continental Magazine, 16, No. 3, Winter 1977, pp. 18–20, illus.
Work of California artist Constance Belda.

NIETO ALCAIDE, VICTOR
"Sobre las vidrieras de la catedral de Sevilla"
Archivo Español de Arte, 48, No. 189, Jan./March 1975, pp. 129–132.

O'BRIEN, VINCENT
Techniques of Stained Glass
New York: Van Nostrand Reinhold Co., 1977, 104 pp., illus.

OLSON, LYNN
"Sculpturing Windows with Cement"
Stained Glass, 72, No. 2, Summer 1977, pp. 92–95, illus.

OLSON, STUART
"Just Color and Something Simple"
Stained Glass, 72, No. 2, Summer 1977, pp. 106–108, illus.
Spokane, Washington, mausoleum.

PERROT, FRANÇOISE
Le vitrail à Rouen
Rouen: Fonds d'Intervention Culturelle et de la Caisse Nationale des Monuments Historiques, 1972, 62 pp., illus

PERROT, FRANÇOISE
"Le vitrail des Chars"
Connaissance des Arts, 310, Dec. 1977, pp. 114–121, illus.

PERROT, FRANÇOISE
"Les vitraux de l'Hôtel-Dieu de Provins XVe–XVIe s.)"
Bulletin de la Société d'histoire et d'archéologie de Provins, No. 128, 1974, pp. 39–51, illus.

PERROT, FRANÇOISE
"Les vitraux du Château d'Écouen. Contribution à l'étude du vitrail civil de la Renaissance"
Actes du Colloque International sur l'Art de Fontainebleau, Oct. 1972, 1975, pp. 175–184, illus.

PEZZELLA, SALVATORE
Il trattato di Antonio da Pisa sulla fabbricazione della vetrate artistiche
Perugia: Umbria Editrice, 1976, 70 pp., illus.

PFAFF, KONRAD
Ludwig Schaffrath; Glasmalerie und Mosaik
Krefeld: Scherpe Verlag, n.d., [1977], 172 pp., illus.

PHILIPPE, ANDRÉ
Le vitrail de la Passion dans l'église de Saint Jean-aux-Bois
Épinal: Imprimerie Pernot, 1921, 14 unbound plates.

POPESCO, PAUL
Die Kathedrale von Chartres
Augsburg: Verlag Josef Hannesschläger, 1969, 144 pp., illus.

PROVIDENCE, RHODE ISLAND. BROWN UNIVERSITY
Transformations of the Court Style; Gothic Art in Europe 1270 to 1330
Providence: Brown University, 1977, 163 pp., illus.

QUAGLIATA, NARCISSUS
"Art in Architecture"
Glass (formerly *Glass Art*), 5, No. 6, Oct. 1977, pp. 26–37, illus.

RAMBUSCH, HAROLD
"An Autobiography"
Stained Glass, 72, No. 2, Summer 1977, pp. 85–88, illus.
Rambusch Decorating Co. owner.

RIGAL, JULIETTE
"Deux dessins de Linard Gontier pour des vitraux de l'Hôtel de l'Arquebuse à Troyes"
La Revue du Louvre et des Musées de France, 27, No. 2, 1977, pp. 70–77, illus.

RIGAN, OTTO
"Glass and Light. A Perfect Marriage"
Los Angeles Times Home Magazine, April 10, 1977, pp. 8–9, illus.

RODE, HERBERT
"Glasmalerei in Köln"
Vor Stefan Lochner: Die Kölner Maler von 1300–1430
Cologne: Wallraf-Richartz-Museum, 1974, pp. 55–58, illus.

SHIPLEY, NAN
"John Allward Stained Glass Windows"
Canadian Collector, 12, No. 1, Jan./Feb. 1977, pp. 43–45, illus.

SKEAT, FRANCIS
"Some Hertfordshire Stained Glass Artists (1)"
Hertfordshire Countryside: The Country Magazine, 31, No. 208, 1976, pp. 16–18, illus.
"Some Hertfordshire Stained Glass Artists (2); The Harpenden Firm Made Some of the Finest Windows"
Hertfordshire Countryside: The Country Magazine, 31, No. 209, 1976, pp. 28–29, illus.
"Some 19th Century Stained Glass Artists and Their Families"
Family History, 19, No. 36, Nov. 1976, pp. 17–55, illus.
London firms of Clayton and Bell, Burlison and Grylls, Charles Clutterbuck.

SPRING, R. O. C.
The Stained Glass of Salisbury Cathedral
Salisbury: The Friends of Salisbury Cathedral, 1973, 22 pp., illus.

Stained Glass Windows of Douglas Strachan
Aberlemno, Scotland: A. C. Russell, 1972, 50 pp.

TAOLIN, ANNA
"The World's Largest Stained Glass Window"
Stained Glass, 72, No. 2, Summer 1977, pp. 99–102, illus.
Resurrection Mausoleum, Justice, Illinois.

TEMME, NORMAN
"Artistry in Glass; Windows and Work of Franklin Art Glass Studios"
Stained Glass, 72, No. 1, Spring 1977, pp. 17–19, illus.
"Louis Comfort Tiffany . . . The Legend Lives On"
Stained Glass, 72, No. 2, Summer 1977, pp. 96–98, illus.
"Sixteen Triptychs Tell Poland's History"
Stained Glass, 72, No. 2, Summer 1977, pp. 103–105, illus.

TETRA, ROOSEVELT
"Catherine Thompson's Oriental Furniture"
Glass (formerly *Glass Art*), 5, No. 6, Oct. 1977, pp. 18–25, illus.

TONEY, KEN
"From Upstart to Establishment: Glassart Studio"
Stained Glass, 72, No. 4, Winter 1977–78, pp. 240–244, illus.

UYTVEN, R. VAN
"Leuvense glasschilders, Klaas Rombouts en Croy-Ramen te Aarschot en elders"
Arca Lovaniensis, 2, 1973, pp. 61–89, illus., English summary.
Early 16th c. Louvain glasspainters.

VAN PESKI, CLAIRE
"Etching on Flashed Glass"
Stained Glass, 72, No. 4, Winter 1977–78, pp. 248–250, illus.

VIDALENC, G.
William Morris
Paris: Librarie Félix Alcan, 1920, 166 pp., illus.

VIGNAU-WILBERG, THEA
"Zur Entstehung zweier Emblemata von Christoph Murer"
Anzeiger des Germanischen Nationalmuseums, 1977, pp. 85–94, illus.
Swiss artist, late 16th c., who made allegorical glass paintings.

VOLLMER, JOYCE
"The *Glass* Report: Lead Came"
Glass (formerly *Glass Art*), 5, No. 5, June 1977, pp. 40–42.

WALLACH, JED
"Stained Glass in France (a Personal View)"
Glass (formerly *Glass Art*), 5, No. 5, June 1977, pp. 10–14, illus.

WASHINGTON, D.C. NATIONAL COLLECTION OF FINE ARTS
". . . and There was Light;" Studies by Abraham Rattner for the Stained Glass Window, Chicago Loop Synagogue
Washington, D.C.: Smithsonian Institution Press, 1976, 32 pp., illus.

WEINBERG, H. BARBARA
The Decorative Work of John La Farge
New York: Garland Publishing, Inc., 1977, 540 pp., illus.

WILKS, HELEN
"Stained Glass at York"
Crafts, 25, March/April 1977, p. 11.
Symposium on the conservation of stained glass.

WILSON, H. WEBER
"How to Repair Leaded Glass"
The Old-House Journal, 3, No. 8, Aug. 1975, pp. 5–8, illus.

Necrology

JEROME STRAUSS, 1893–1978

To a professional, in museums or elsewhere, the word "amateur" conjures an image of superficiality, of well-intentioned meddlesomeness, or of buoyant but uncritical eagerness. Indeed, this may well describe certain individuals. But the description belies the roots of a word that comes from "to love," and loving is understanding, is patience, is sharing, is communicating not for gratification of the self, but for a deeper understanding and to provide a truer meaning to life. It is in this sense that Jerome Strauss was an "amateur."

His distinctions were many, and his knowledge was vast. An experienced professional metallurgist, inventor, and businessman, these aspects of his life alone would have gained him lasting recognition. Through his knowledge of the complex chemistry and structure of rare metals, he contributed richly to his country's defense, to the ascendency of American steel, and to the prosperity of the company with whom he was associated for over fifty years, the Vanadium Corporation of America. Sure of his facts, firm in communicating them, Jerome Strauss must have exuded in the boardroom the same calm confidence and patience that he later projected in the lecture hall or at the auction house.

Others who knew his earlier career better than I may relate what first prompted him to collect, but collect he did and from an early age. Attracted by the intricacy and the fragility of glass, perhaps because it was so different from the other products of the fire upon which his life was centered, in the early 1930's he started acquiring pressed glass and did so with such avidity and thoroughness that he was able to discuss his holdings and their meaning on an equal footing with the best glass collectors and scholars of the period.

But for a cultivated mind, well-read, fluent in French and German, and used to travel abroad, pressed glass was too limited a field. The collection of early American pressed glass was sold, and he embarked upon the search which was to continue for the rest of his life: the pursuit of the glass drinking form from its very earliest inception down to the most recent of contemporary variants. He sought the form, not in its largest or more spectacular variety, but through its infinite permutations, from Egypt to Mesopotamia, Syria to Rome, China to England, ferreting out shape and substance with encyclopedic thoroughness and an unswerving sense of inquiry.

The substance, its origins, methods of manufacture, the influences that it reflected and the legacies that it fostered were all studied. And thus in a half century was developed one of the most distinguished, thoroughly documented, and varied collections on the subject ever assembled by one man.

The collection by itself would have been an impressive accomplishment and an unmatchable legacy, but Jerome Strauss was also concerned with the other facets relating to his subject. He espoused the principle that if a glass was worth acquiring, the print, document, manuscript, and book that illuminated its origins or reflected its influences were of equal importance. Like the professional antiquarian, he studied sources, consulted with colleagues, visited museums and shops, constantly seeking new relationships and gaining more profound insights. In this process he became the friend of virtually every glass scholar of note and the client of virtually every major glass dealer.

Strauss was not one of those collectors who relished in the spectacular; he glowed in the significant. Being a scientist, he turned to chemical analysis and carried out the most careful "autopsies" to understand what made an object tick. He was among those who did not dismiss fakes since, through the study of fakes, he knew that one could learn much about those elements that reflect genuineness. He was not obsessed by "proof condition," and the damaged testimonies of past creativity were acquired and studied with the same reverence and thoroughness as the most spectacular intact object.

The acquisition of objects and the pursuit of knowledge can be dry, humorless, obsessive, but for Jerome they became a means for expanding human contacts, discovering new societies, and, eventually, buffering the grief of the widower.

An indefatigable lecturer, a pillar of The Corning Museum of Glass Seminars, he was equally relentless in the pursuit of the document or the refinement of the footnote. His sagacious eye and critical comments contributed much to this *Journal* and to other learned publications.

It is no coincidence that he was equally concerned about the beverages that many of his glasses were made to contain. An oenophile of note, he savored the bouquet, enjoyed the color as it played in the transparent glass, adding to the act of imbibing the touch of elegance, scholarship, and restraint that was revealed in all his activities. To visit a museum, hear the lecture of a distinguished colleague, examine the wear marks on a base or a lip, to read the results of new archaeological discoveries or about the acquisition of an important object by a fellow collector or institution, had a special meaning in his company, for he always had the *mot juste* and it was always said with kindness and appreciation.

The collection that he formed and which he named in honor of his late wife, Ruth Bryan Strauss, and to which he added until the very end of his life, is his tactile legacy. But for one who knew the twinkle in his eye, who witnessed the affectionate and considerate bonds that linked him to the companion of his later years, Lucille Jackson Strauss, who relied heavily on his advice in sharpening his own connoisseurship, and who had the rare pleasure of sharing the delicate bouquet of his cellar, it is the man who will be missed: a true and generous friend, an exemplary "amateur."

Paul N. Perrot
October 1, 1978

DR. PETER LA BAUME

Die Historischen Museen der Stadt Köln beklagen den jähen Tod von Herrn Hauptkustos Dr. Peter La Baume, Abteilungsdirektor am Römisch-Germanischen Museum.

Er erlag am 7. 12. 1977 einer tückischen Krankheit. Seit Jahrzehnten hatte seine ganze Liebe dem römischen Kunstgewerbe gegolten, dem er neben vielen Einzelabhandlungen eine grundlegende Darstellung gewidmet hat, die noch lange von seiner hohen Kennerschaft zeugen wird. In den letzten Jahren hatte er sich durch intensive Studien auf eine neue Behandlung der römischen Glasmanufakturen in der römischen Welt vorbereitet. Dabei war es sein Ziel, Kriterien für die längst überfällige neue Behandlung der Gläser aus den Kölner Produktionsstätten zu gewinnen. So sehr diese nämlich inmitten der Gesamtproduktion römischer Gläser im römischen Reich gesehen werden müssen, so scharf lassen sie sich doch als eine eigene Gruppe Kölner Gläser aussondern und würdigen. Mit ganz besonderer Liebe hat Dr. Peter La Baume seine ganze Sachkenntnis diesem Problemkreis zugewandt und dabei neue Beurteilungskriterien entwickelt, die in der ganzen Breite ihm nicht mehr anzuwenden vergönnt war. Die Sachforschung der Kölner Archäologie hat damit einen schwer zu verwindenden Verlust erlitten.

Immer wieder hat sich Dr. La Baume in den 24 Jahren seiner Tätigkeit am Römisch-Germanischen Museum zu den Problemen der römischen Frühgeschichte kritisch und anregend geäussert. Baustein und Baustein hat er mit der ihm eigenen Zähigkeit zusammengetragen. Eine schöne Nüchternheit war dabei seine Leitlinie. Das hinderte ihn an der Überschätzung der ihm zur Übertragung zugesprochenen Fundkomplexe. Diese Grundhaltung sicherte seinen Aussgaen nicht allein Ausgewogenheit, sondern wird bewirken, dass seine überlegen abgesicherten Äusserungen noch in weiter Zukunft als solide Bausteine für die Kenntnis vom römischen Köln ohne Vorbehalt benutzt werden können.

Das neue Römisch-Germanische Museum trägt mit seine Züge. Sein Beitrag dazu ist die klare Darbietung der Sachgüter in den Vitrinen. Er hat in langen und anregenden Diskussionen vor jedem Überschwang gewarnt. Seine grosse Leistung inmitten dieser Gesamtinszenierung ist und bleibt die Darbietung des kölnischen Prunkglases. Jedes Stück dieses Museumshöhepunktes hat er selbst aus den reichen Beständen abwägend augesucht, mit Bedacht das eine zum anderen Stück geordnet. Hier gestattete er seiner tiefen Liebe zu den schönen Gläsern alle Hingabe deren er fähig war. Dies spiegelt die überzeugende Ordnung der granzilen Versammlung.

Inmitten der Museumsfamilie bewährte sich Dr. Peter La Baume als Mensch. Er war die gute Seele des Hauses, immer Ausgleich bedacht, allen Sorgen allen Mitarbeitern gegenüber offen, leise, behutsam, nie auf sich, immer auf andere bedacht. Nicht nur in der wissenschaftlichen Erforschung des römischen und fränkischen Köln hinterlässt er aufgrund seines breiten und soliden Wissens, sondern vor allem als Mensch eine grosse Lücke, die kaum zu schliessen sein wird. Nichts kann dies deutlicher belegen, als die Feststellung aller Mitarbeiter der Historischen Museen der Stadt Köln: Die Trauer über seinen jähen Tod ist gross.

DR. HUGO BORGER, *Director*
Historische Museen
Museen der Stadt Köln

CONTRIBUTORS

ROBERT H. BRILL, Research Scientist, The Corning Museum of Glass; formerly Assistant Professor of Chemistry, Upsala College.

ROBERT J. CHARLESTON, formerly Keeper, Department of Ceramics, Victoria and Albert Museum, London. (retired)

HELMUT RICKE, Deputy Director and Curator for Sculpture, Decorative Arts and Glass, Kunstmuseum Düsseldorf, Düsseldorf.

The design, composition, and printing of this book
are by The Stinehour Press, Lunenburg, Vermont,
and The Meriden Gravure Company, Meriden, Connecticut.

The binding is by The Stinehour Press.